Introduction to Optimization Meth and their Application in Statistics

OTHER STATISTICS TEXTS FROM
CHAPMAN AND HALL

The Analysis of Time Series
C. Chatfield

Statistics for Technology
C. Chatfield

Applied Statistics
D. R. Cox and E. J. Snell

Introduction to Multivariate Analysis
C. Chatfield and A. J. Collins

An Introduction to Statistical Modelling
A. J. Dobson

Multivariate Statistics–A Practical Approach
B. Flury and H. Riedwyl

Multivariate Analysis of Variance and Repeated Measures
D. J. Hand and C. C. Taylor

Multivariate Statistical Methods – a primer
Bryan F. Manley

Statistical Methods in Agriculture and Experimental Biology
R. Mead and R. N. Curnow

Elements of Simulation
B. J. T. Morgan

Essential Statistics
D. G. Rees

Decision Analysis: A Bayesian Approach
J. Q. Smith

Applied Statistics: A Handbook of BMDP Analyses
E. J. Snell

Elementary Applications of Probability Theory
H. C. Tuckwell

Intermediate Statistical Methods
G. B. Wetherill

Further information on the complete range of Chapman and Hall *statistics books is available from the publishers.*

Introduction to Optimization Methods and their Application in Statistics

B. S. Everitt BSc MSc

Reader in Statistics in Behavioural Science and
Head of Biometrics Unit, Institute of Psychiatry

London New York
CHAPMAN AND HALL

First published in 1987 by Chapman and Hall Ltd
11 New Fetter Lane, London EC4P 4EE
Published in the USA by Chapman and Hall
29 West 35th Street, New York NY 10001

© 1987 B. S. Everitt

Printed in Great Britain by St Edmundsbury Press Ltd
Bury St Edmunds, Suffolk

ISBN 0 412 272105 (HB)
ISBN 0 412 296802 (PB)

This title is available in both hardbound and paperback editions. The paperback edition is sold subject to the condition that it shall not, by way of trade or otherwise, be lent, resold, hired out, or otherwise circulated without the publisher's prior consent in any form of binding or cover other than that in which it is published and without a similar condition including this condition being imposed on the subsequent purchaser.

All rights reserved. No part of this book may be reprinted, or reproduced or utilized in any form or by any electronic, mechanical or other means, now known or hereafter invented, including photocopying and recording, or in any information storage and retrieval system, without permission in writing from the publisher.

British Library Cataloguing in Publication Data

Everitt, Brian
Introduction to optimization methods and
their applications in statistics.
1. Mathematical statistics 2. Mathematical
optimization
I. Title
519.5 QA276.A1

ISBN 0-412-27210-5
ISBN 0-412-29680-2 Pbk

Library of Congress Cataloging in Publication Data

Everitt, Brian.
Introduction to optimization methods and their
application in statistics.

Bibliography: p.
Includes index.
1. Mathematical statistics. 2. Mathematical
optimization I. Title.
QA276.E92 1987 519.5 87-11693
ISBN 0-412-27210-5
ISBN 0-412-29680-2 (pbk.)

Contents

Preface

Optimization techniques are used to find the values of a set of parameters which maximize or minimize some objective function of interest. Such methods have become of great importance in statistics for estimation, model fitting, etc. This text attempts to give a brief introduction to optimization methods and their use in several important areas of statistics. It does not pretend to provide either a complete treatment of optimization techniques or a comprehensive review of their application in statistics; such a review would, of course, require a volume several orders of magnitude larger than this since almost every issue of every statistics journal contains one or other paper which involves the application of an optimization method.

It is hoped that the text will be useful to students on applied statistics courses and to researchers needing to use optimization techniques in a statistical context.

Lastly, my thanks are due to Bertha Lakey for typing the manuscript.

B. S. Everitt
August 1986

1

An introduction to optimization methods

1.1 INTRODUCTION

A problem considered in all basic statistics courses is that of finding estimates of the two parameters in a simple linear regression model relating a dependent variable, y, to an explanatory variable, x. The model is usually formulated as

$$y_i = \alpha + \beta x_i + \epsilon_i \tag{1.1}$$

where $x_i, y_i, i = 1, \ldots, n$ are the values of the explanatory and dependent variable for a sample of observations considered to arise from the model, and the $\epsilon_i, i = 1, \ldots, n$ are 'error' or residual terms with zero expected values, accounting for how much an observation, y_i, differs from its predicted value, $\alpha + \beta x_i$.

The problem of finding estimates of the parameters, α and β, of the regression model in (1.1) may be approached in several ways. Perhaps the most common is to seek some goodness-of-fit criterion which measures, in some sense, how closely the model agrees with the observed data, and then choose values for the two parameters which *minimize* the chosen measure of fit. An obvious goodness-of-fit criterion for the simple linear regression model is the sum-of-squares of the error terms in (1.1), that is

$$S = \sum_{i=1}^{n} \epsilon_i^2 \tag{1.2}$$

Clearly S does measure how well the observed values of the dependent variable fit those predicted by the model, with smaller values of S indicating a better fit. Consequently choosing as estimates of α and β those values which minimize S is an intuitively reasonable procedure and is, of course, nothing less than the well-known *least squares* estimation technique.

Another commonly occurring estimation problem in statistics arises when we wish to estimate the parameter or parameters of a probability density function given a random sample taken from the density function. For example, we may have a sample of n values, $x_1, \ldots, x_n$, from an exponential density function of the form

$$f(x) = \lambda e^{-\lambda x} \qquad x>0 \tag{1.3}$$

and we wish to estimate λ. A very useful estimation procedure in this situation is to form the joint probability density function of the observations, that is

$$\mathscr{L}(x_1, \ldots, x_n; \lambda) = \prod_{i=1}^{n} \lambda e^{-\lambda x_i} \tag{1.4}$$

and choose as the estimate of λ the value which maximizes $\mathscr{L}$, which is generally referred to as the *likelihood function*. This procedure will also be well known to most readers as *maximum likelihood estimation*.

Both the estimation problems described above can be formulated in terms of *optimizing* some numerical function with respect to a number of parameters, and many other statistical problems may be formulated in a similar manner. It is methods for performing such optimizations and their application in statistics which are the main concern of this text.

1.2 THE OPTIMIZATION PROBLEM

In its most general form the problem with which we will be concerned involves finding the optimum value (maximum or minimum) of a function $f(\theta_1, \ldots, \theta_m)$ of m parameters, $\theta_1, \ldots, \theta_m$. We should note at this stage that from a mathematical point of view there is little point in considering both maximization and minimization since maximizing f is equivalent to minimizing $-f$; consequently the discussion in the remainder of the text will normally be confined to minimization. The values taken by the parameters may in some situations be *constrained* and in others *unconstrained*. For example, in the linear regression model of the previous section, the parameters α and β may both take any real value; in other words, they are unconstrained. The parameter of the exponential distribution in (1.3) is, however, constrained to take only positive values. Some comments about the constrained optimization problem will be made in Section 1.5.

Many of the concepts we shall need in discussing optimization methods can be introduced via the case of a function of a single parameter, and Fig. 1.1 shows a graphical representation of such a function. This graph shows that the function has two minima, one at θ_0 and one at θ_1, a maximum at θ_2, and a point of inflexion at θ_3. The minimum at θ_0 is known as a *local minimum* since the value of $f(\theta_0)$ is lower than $f(\theta)$ for values of θ in the neighbourhood of θ_0; the minimum at θ_1 is known as a *global minimum* since $f(\theta_1)$ is lower than $f(\theta)$ for *all* values of θ. As we shall see later, a major problem in complex minimization problems is to decide whether we have found a local or global minimum.

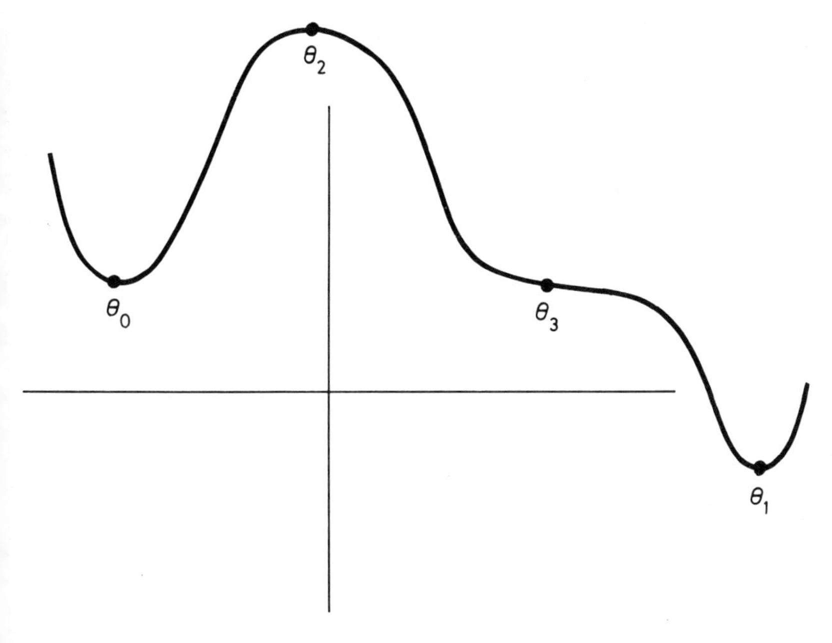

Figure 1.1 Function of a single parameter showing a maximum (θ_2), a local minimum (θ_0), global minimum (θ_1) and point of inflexion (θ_3).

The classical approach to the problem of finding the values θ_0 and θ_1 is to note that at both θ_0 and θ_1 the gradient of $f(\theta)$ is zero, so that θ_0 and θ_1 will be solutions of the equation

$$\frac{df}{d\theta} = 0 \tag{1.5}$$

As we can see from Fig. 1.1 the value θ_2, at which there is a local maximum, and θ_3, at which there is a horizontal point of inflexion, also satisfy this equation; consequently satisfying equation (1.5) is a *necessary* but not a *sufficient* condition for a point to be a minimum. However, examining again Fig. 1.1, we see that at θ_0 and θ_1 the gradient changes sign from negative to positive, at θ_2 the change is from positive to negative, and at θ_3 the gradient does not change sign. So at a minimum the gradient is an increasing function; the rate of change of the gradient is measured by the second derivative so for a minimum we require

$$\frac{d^2f}{d\theta^2} > 0 \tag{1.6}$$

when evaluated at the suspected minimum point.

These ideas may be extended to the minimization of a function of several variables, $f(\theta_1, \ldots, \theta_m)$, so that a necessary condition for a minimum is that

$$\frac{\partial f}{\partial \theta_1} = \frac{\partial f}{\partial \theta_2} = \ldots = \frac{\partial f}{\partial \theta_m} = 0 \tag{1.7}$$

Solutions to these equations may also represent maxima or saddle points, and these various possibilities are illustrated for a function of two variables by the contour diagram shown in Fig. 1.2.

On this diagram, P_1, P_2, P_3 and P_4 are points at which equations (1.7) are satisfied; the corresponding values of f are 0.0, 2.5, 6.5 and 3.5. P_1 is the *global minimum*, that is the required overall minimum of the function. P_2 is a *local minimum*, that is $f(P_2)$ is less than f for all points in the immediate neighbourhood of P_2 but $f(P_2) > f(P_1)$. P_3 is a *local maximum* of f, and P_4 is a *saddle point*; along the direction AB it corresponds to a maximum of f, while along CD it corresponds to a minimum.

The *sufficient* condition for a solution of (1.7) to be a minimum, corresponding to the requirement given in (1.6) for the single-parameter case, is that the matrix **H** with elements h_{ij} given by

$$h_{ij} = \frac{\partial^2 f}{\partial \theta_i \partial \theta_j} \tag{1.8}$$

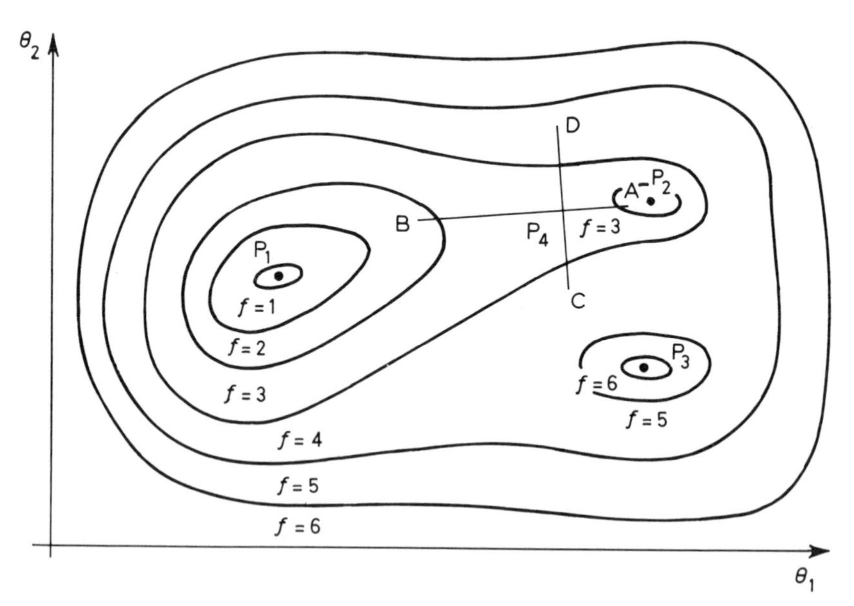

Figure 1.2 Contour diagram of a function of two parameters having a local maximum, a local minimum and a saddle point.

be positive definite when evaluated at the point being considered. **H** is known as the *Hessian* matrix; it is symmetric and of order $m \times m$.

1.3 SOME SIMPLE EXAMPLES

Let us return to the two examples described in Section 1.1 to illustrate a number of the points made in Section 1.2. Consider first the problem of the maximum likelihood estimation of the parameter of an exponential density function. The likelihood function for a sample of n values is given by

$$\mathscr{L}(x_i, \ldots, x_n; \lambda) = \prod_{i=1}^{n} \lambda e^{-\lambda x_i} \tag{1.9}$$

We wish to choose that value of λ which maximizes $\mathscr{L}$ or, equivalently minimizes $-\mathscr{L}$. As with most maximum likelihood problems a simplification is achieved if we consider not $\mathscr{L}$ but the log-likelihood function, L, given by

$$L = \log_e(\mathscr{L}) \tag{1.10}$$

$\mathscr{L}$ and L clearly have their maxima at the same points but in practice L is usually far more convenient to deal with. For the exponential density

$$L = n \log_e \lambda - \lambda \sum_{i=1}^{n} x_i \tag{1.11}$$

Consequently the function we require to minimize is

$$F = \lambda \Sigma x_i - n \log_e \lambda \tag{1.12}$$

Differentiating with respect to λ gives

$$\frac{\mathrm{d}F}{\mathrm{d}\lambda} = \Sigma x_i - \frac{n}{\lambda} \tag{1.13}$$

Setting $\mathrm{d}F/\mathrm{d}\lambda$ to zero leads to the following estimator for λ:

$$\hat{\lambda} = n / \sum_{i=1}^{n} x_i \tag{1.14}$$

that is the reciprocal of the sample mean. Clearly this corresponds to a minimum of F since

$$\frac{\mathrm{d}^2F}{\mathrm{d}\lambda^2} = \frac{n}{\lambda^2} \tag{1.15}$$

which is always positive.

Now let us consider the least squares estimation of the two parameters in the simple linear regression model. This involves minimization of the goodness-of-fit criterion specified in (1.2), which may be rewritten as follows:

$$S = \sum_{i=1}^{n} (y_i - \alpha - \beta x_i)^2 \tag{1.16}$$

so that the equations given by (1.7) take the form

$$\frac{\partial S}{\partial \alpha} = -2 \sum_{i=1}^{n} (y_i - \alpha - \beta x_i) = 0, \tag{1.17}$$

$$\frac{\partial S}{\partial \beta} = -2 \sum_{i=1}^{n} x_i(y_i - \alpha - \beta x_i) = 0 \tag{1.18}$$

Solving these two equations leads to the following well-known estimators for α and β:

$$\hat{\alpha} = \bar{y} - \hat{\beta}\bar{x}, \tag{1.19}$$

$$\hat{\beta} = \frac{Cxy}{Cxx}, \tag{1.20}$$

where

$$Cxy = \sum_{i=1}^{n} x_i y_i - \sum_{i=1}^{n} x_i \sum_{i=1}^{n} y_i / n, \tag{1.21}$$

$$Cxx = \sum_{i=1}^{n} x_i^2 - \left(\sum_{i=1}^{n} x_i \right)^2 / n, \tag{1.22}$$

The Hessian matrix for this problem is given by

$$\mathbf{H} = \begin{bmatrix} \dfrac{\partial^2 S}{\partial \alpha^2} & \dfrac{\partial^2 S}{\partial \alpha \, \partial \beta} \\ \dfrac{\partial^2 S}{\partial \beta \, \partial \alpha} & \dfrac{\partial^2 S}{\partial \beta^2} \end{bmatrix}, \tag{1.23}$$

$$= \begin{bmatrix} 2n & 2\sum_{i=1}^{n} x_i \\ 2\sum_{i=1}^{n} x_i & 2\sum_{i=1}^{n} x_i^2 \end{bmatrix} \tag{1.24}$$

It is easy to show that **H** is positive definite and consequently that (1.19) and (1.20) correspond to a minimum of the function S.

In both these cases the solutions to the equations specifying the minimum (equations (1.5) and (1.7)) could be solved directly to give estimators for the parameters which were simple functions of the observations. In many situations, however, these equations cannot be solved directly, and other approaches must be adopted to the minimization problem. Some general characteristic of the type of procedure necessary are discussed in the following section; detailed descriptions of specific techniques will be left until Chapters 2 and 3.

1.4 MINIMIZATION PROCEDURES

The minimization techniques to be discussed in the next two chapters all have certain features in common. The most obvious is that they are *iterative* and proceed by generating a sequence of solutions each of which represents an improved approximation to the parameter values at the minimum of f in the sense that

$$f(\boldsymbol{\theta}_{i+1}) \leq f(\boldsymbol{\theta}_i) \tag{1.25}$$

where $\boldsymbol{\theta}_{i+1}$ and $\boldsymbol{\theta}_i$ are vectors containing the values of the m parameters at iterations $i+1$ and i. Such procedures require an initial set of parameter values, $\boldsymbol{\theta}_0$, generally supplied by the investigator, from which successive approximations arise by means of an equation of the form

$$\boldsymbol{\theta}_{i+1} = \boldsymbol{\theta}_i + h_i \mathbf{d}_i. \tag{1.26}$$

In this equation $\mathbf{d}_i$ is an m-dimensional vector specifying the *direction* to be taken in moving from $\boldsymbol{\theta}_i$ to $\boldsymbol{\theta}_{i+1}$ and h_i is a scalar specifying the *distance* to be moved along this direction.

The choice of a suitable direction and distance (often referred to as the *step size*) to ensure that (1.25) is satisfied may be made in a number of ways; it may rely solely on values of the function plus information gained from earlier iterations, or on values of the partial derivatives of f with respect to the parameters. Techniques adopting the first approach are generally known as *direct search methods* and are discussed in Chapter 2. The second type of approach, *gradient methods* are the subject of Chapter 3.

A problem common to all the techniques to be discussed in the next two chapters is how to decide when the iterative procedure has reached the required minimum. In general such decisions are taken on the basis of the sequences $\{\boldsymbol{\theta}_i\}$ and $\{f(\boldsymbol{\theta}_i)\}$, and possible convergence criteria are

$$|f(\boldsymbol{\theta}_{i+1}) - f(\boldsymbol{\theta}_i)| < \epsilon \tag{1.27}$$

and/or

$$||\boldsymbol{\theta}_{i+1}-\boldsymbol{\theta}_i||<\epsilon' \quad (1.28)$$

for prescribed values of ϵ and ϵ'. Although such criteria are commonly used and are, in many situations, satisfactory, they can in some circumstances cause the iterative procedure to be terminated prematurely. For example, Fig. 1.3 illustrates a case where terminating the iterations on the basis of the fractional changes in $f(\boldsymbol{\theta})$ being less than some small number, causes the procedure to finish on a flat plateau. Figure 1.4 illustrates a case where the use of (1.28) causes premature termination on a very steep slope.

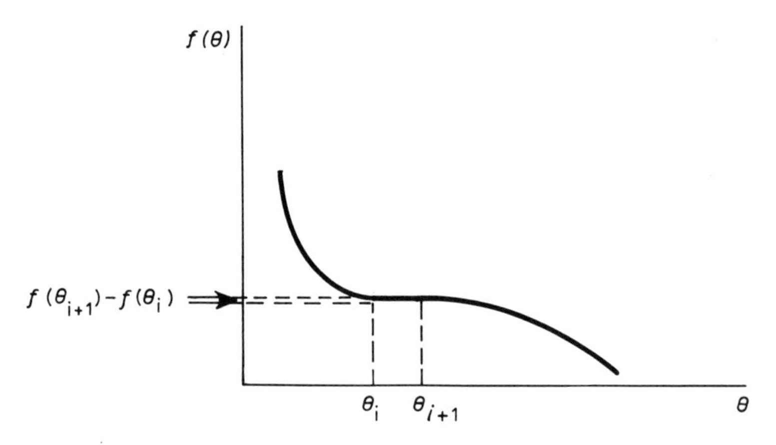

Figure 1.3 Premature termination on a flat plateau.

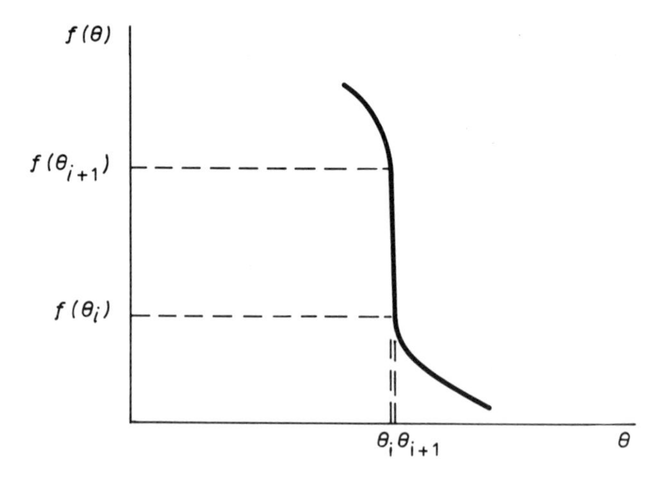

Figure 1.4 Premature termination on a steep slope.

In order to guard against such possibilities a more stringent convergence criterion might be used in which (1.27) or (1.28) were required to hold for each of a number of consecutive iterations.

Termination criteria such as (1.27) and (1.28) are strongly dependent on the scaling of both the objective function, f, and the parameters, $\theta_1, \ldots, \theta_m$. For example, if $\epsilon = 10^{-3}$ and f is always in the interval $(10^{-7}, 10^{-5})$, then it is likely that any values of $\theta_1, \ldots, \theta_m$ will satisfy (1.27). A problem arises with (1.28) if the parameters are on very different scales, since if $m = 2$, and θ_1 is in the range (10, 100) and θ_2 in the range (0.001, 0.01), then (1.28) will virtually ignore the second parameter. This problem of scale will also affect those optimization methods which are not invariant with respect to scale changes. The obvious solution to this problem is to choose units for the parameter so that each has roughly the same magnitude. For more detailed comments about stopping criteria and scaling see Dennis and Schnabel (1983, Ch. 7).

1.5 CONSTRAINED MINIMIZATION

In the discussion in previous sections it has been implicitly assumed that the elements of the parameter vector, $\boldsymbol{\theta}$, are not subject to any constraints. This is not always the case, however, and problems do arise where we wish to minimize some objective function, $f(\boldsymbol{\theta})$, subject to various constraints on the parameters. Such constraints may be equalities, for example,

$$\theta_1^2 + \theta_2^2 + \theta_3^2 = 1 \tag{1.29}$$

or inequalities,

$$\theta_1 + \theta_2 + \theta_3 > 0 \tag{1.30}$$

Constraints on the parameters in statistical problems may arise for a number of reasons; the parameters may, for example, be variances which must be greater than zero, or proportions which must lie between zero and one.

The simplest method of dealing with constrained optimization problems is to reparametrize so that they become unconstrained. For example, if an original parameter is subject to constraints of the form

$$\theta_i > c_i \tag{1.31}$$

$$a_i < \theta_i < b_i \tag{1.32}$$

where a_i, b_i and c_i are constants, then defining a new parameter α_i as

$$\alpha_i^2 = \theta_i - c_i \tag{1.33}$$

$$\sin^2\alpha_i = (\theta_i - a_i)/(b_i - a_i) \tag{1.34}$$

removes the constraints (1.31) and (1.32) and allows an unconstrained optimization for the parameter α_i. Particularly common in statistics is the situation where a_i and b_i in (1.32) are zero and unity and the parameter θ represents a proportion or probability. A commonly used transformation in this case is the *logistic*,

$$\alpha_i = \log \frac{\theta_i}{1-\theta_i} \tag{1.35}$$

More formal methods of dealing with constrained optimization problems such as *Lagrange multipliers* and *penalty functions* are described in Rao (1979 Ch. 7). It is important to emphasize, however, that many problems with simple constraints can be solved by unconstrained algorithms, because the constraints are satisfied by the unconstrained minimizer.

1.6 SUMMARY

Many problems in statistics may be formulated in terms of the minimization of some function with respect to a number of parameters. In most cases the equations for a minimum arising from (1.7) cannot be solved directly, and iterative procedures are needed. During the last two decades there have been major advances in such techniques and this has had considerable impact in many branches of statistics, as we shall attempt to describe in later chapters. The next two chapters concentrate on describing a number of commonly used minimization methods. It should be emphasized that they do not attempt to provide a comprehensive account of such techniques, only to provide a basis for a discussion of their use in a statistical context. Many excellent *detailed* accounts of optimization methods are available elsewhere, for example Bunday (1984), Walsh (1975) and Rao (1979).

2

Direct search methods

.1 INTRODUCTION

ı this chapter we shall describe a number of *direct search methods* for ıinimization. Such methods do *not* require the explicit evaluation of any artial derivatives of the function being minimized, but instead rely solely on alues of the function found during the iterative process. In some cases these ınction values are used to obtain numerical approximations to the derivaves of the objective function, in others they provide the basis for fitting w-order polynomials or surfaces to the function in the vicinity of the inimum. We first consider the minimization of a function of a single ırameter, and then the multiparameter situation.

2 UNIVARIATE SEARCH METHODS

arch methods for minimizing a function of a single variable fall into two asses: those which specify an interval in which the minimum lies and those hich specify the position of the minimum by a point approximating to it. In der to apply the former we shall assume that an initial interval known to ntain the minimum is given and that the function is unimodal within this terval. With such methods we literally search for the minimum of the nction in some interval $a<\theta<b$ by evaluating the function at chosen points the interval. The alternative approach is to use a few function values aluated at particular points to approximate the function by a simple lynomial, at least over a limited range of values. The position of the ıction minimum is then approximated by the position of the polynomial nimum, the latter being relatively simple to calculate. We begin with an ample of the first approach followed by one of the second.

.1 Fibonacci search

 suppose that the required minimum is known to be within the interval $, \theta_2)$, and that two points, θ_3 and θ_4, are to be chosen within this interval so ıt

$$\theta_1<\theta_3<\theta_4<\theta_2 \qquad (2.1)$$

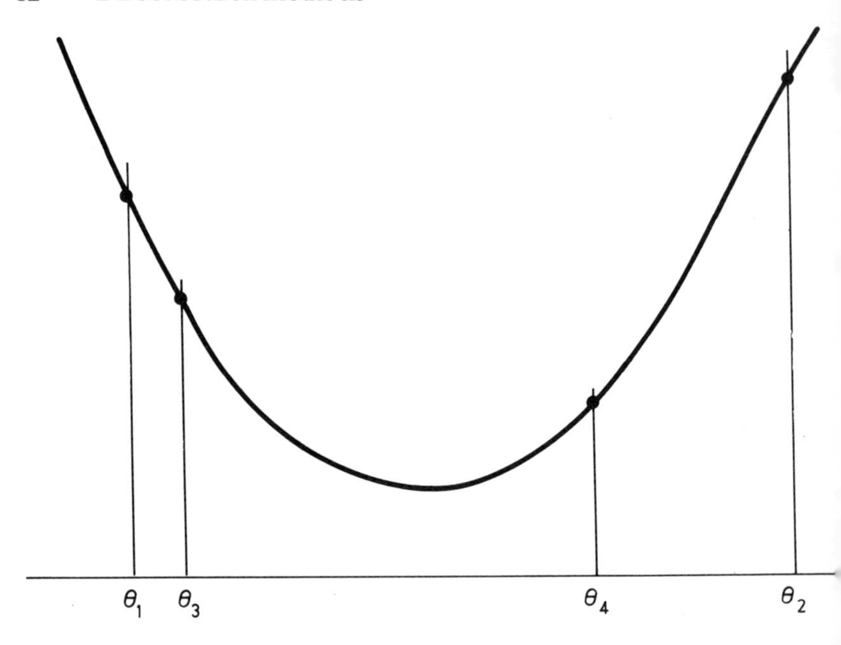

Figure 2.1 Searching for a minimum of a function of a single variable.

Since we are assuming that the function is unimodal in (θ_1, θ_2), it is clear tha $f(\theta_3) \geq f(\theta_4)$ then the minimum lies in (θ_3, θ_2), whereas if $f(\theta_3) \leq f(\theta_4)$, t minimum lies in (θ_1, θ_4). Figure 2.1 illustrates the former situation. Furth reduction of the interval containing the minimum is achieved by evaluati the function at other points within the interval currently being considered.

The question arises as to how to choose the values of θ at which to evalu the function. Clearly we should not make the decision for all points *a prio* but instead let the function values obtained earlier determine the position subsequent points. If we specify that we can only afford to evaluate function a particular number of times, n say, then it can be shown (see Wal 1975) that the most efficient search procedure is that known as the *Fibona* search (efficiency in this context means that no other method can guaran such a large reduction in the search interval with n function evaluations). T procedure uses a sequence of positive integers known as Fibonacci numbe defined by the following equations:

$$F_0 = F_1 = 1 \quad (2$$

$$F_i = F_{i-1} + F_{i-2} \qquad n \geq 2 \quad (2$$

so the sequence begins 1, 1, 2, 3, 5, 8, 13, 21, ... If we assume that $(\theta_1^{(1)}, \theta$

the initial interval to be searched, then the first step of the Fibonacci rocedure is to find

$$I_1 = \frac{F_{n-1}}{F_n} I_0 \tag{2.4}$$

here $I_0 = \theta_2^{(1)} - \theta_1^{(1)}$ is the length of the initial search interval. The two values f θ defining the next interval are then taken as

$$\theta_3^{(1)} = \theta_1^{(1)} + I_1, \tag{2.5}$$

$$\theta_4^{(1)} = \theta_2^{(1)} - I_1, \tag{2.6}$$

$$= \theta_1^{(1)} + \frac{F_{n-2}}{F_n} I_0. \tag{2.7}$$

Some part of this interval is now discarded by using the unimodality sumption, and the process is essentially repeated in the new interval; for tails see Adby and Dempster (1974). By examining the ratio (I_n/I_0) we can

Table 2.1

Value of n	Fibonacci number F_n	Interval reduction ratio
0	1	1.0
1	1	1.0
2	2	0.5
3	3	0.3333
4	5	0.2
5	8	0.1250
6	13	0.07692
7	21	0.04762
8	34	0.02941
9	55	0.01818
10	89	0.01124
11	144	0.006944
12	233	0.004292
13	377	0.002653
14	610	0.001639
15	987	0.001013
16	1597	0.0006406
17	2584	0.0003870
18	4181	0.0002392
19	6765	0.0001479
20	10 946	0.00009135

determine the required number of steps needed to obtain any desire accuracy in locating the optimum point. Table 2.1 gives the reduction ratio i the size of the search interval for different values of n.

An alternative to the Fibonacci search procedure which does not requir that the number of function evaluations be specified in advance is the *golde section method*. In the Fibonacci procedure the location of the first two point is determined by the total number of function evaluations to be performe With the golden section we start with the assumption that we are going t make a large number of function evaluations, the actual number bein decided during the computation. Again details are given in Adby an Dempster (1974).

2.2.2 Quadratic interpolation

If we know the values of a function $f(\theta)$ at three distinct points θ_1, θ_2 and θ then we can approximate $f(\theta)$ by the quadratic function

$$h(\theta) = A\theta^2 + B\theta + C \qquad (2.$$

where A, B and C can be determined from the equations

$$A\theta_1^2 + B\theta_1 + C = f(\theta_1) \qquad (2.$$

$$A\theta_2^2 + B\theta_2 + C = f(\theta_2) \qquad (2.1$$

$$A\theta_3^2 + B\theta_3 + C = f(\theta_3) \qquad (2.1$$

Solution of these equations gives

$$A = -\left[\frac{(\theta_2-\theta_3)f(\theta_1)+(\theta_3-\theta_1)f(\theta_2)+(\theta_1-\theta_2)f(\theta_3)}{(\theta_1-\theta_2)(\theta_2-\theta_3)(\theta_3-\theta_1)}\right] \qquad (2.1$$

$$B = \left[\frac{(\theta_2^2-\theta_3^2)f(\theta_1)+(\theta_3^2-\theta_1^2)f(\theta_2)+(\theta_1^2-\theta_2^2)f(\theta_3)}{(\theta_1-\theta_2)(\theta_2-\theta_3)(\theta_3-\theta_1)}\right] \qquad (2.1$$

$$C = \left[\frac{\theta_2\theta_3(\theta_3-\theta_2)f(\theta_1)+\theta_3\theta_1(\theta_1-\theta_3)f(\theta_2)+\theta_1\theta_2(\theta_2-\theta_1)f(\theta_3)}{(\theta_1-\theta_2)(\theta_2-\theta_3)(\theta_3-\theta_1)}\right] \qquad (2.1$$

Now if we differentiate (2.8) with respect to θ we obtain

$$\frac{dh(\theta)}{d\theta} = 2A\theta + B \qquad (2.1$$

;iving the position of the minimum of h as

$$\theta_{\min} = -\frac{B}{2A} \tag{2.16}$$

vhich using (2.12) and (2.13) gives

$$\theta_{\min} = \frac{1}{2}\left[\frac{(\theta_2^2-\theta_3^2)f(\theta_1)+(\theta_3^2-\theta_1^2)f(\theta_2)+(\theta_1^2-\theta_2^2)f(\theta_3)}{(\theta_2-\theta_3)f(\theta_1)+(\theta_3-\theta_1)f(\theta_2)+(\theta_1-\theta_2)f(\theta_3)}\right] \tag{2.17}$$

assuming $A > 0$ so that $\mathrm{d}^2h/\mathrm{d}\theta^2 > 0$).

The value given by equation (2.17) acts as an approximation to the required ninimum of $f(\theta)$.

To use this approach in practice we assume we have an initial approxim-ation to the position of the minimum, θ^*, and a preselected trial step length, l. Ve then begin with the points

$$\theta_1 = \theta^*, \qquad \theta_2 = \theta^* + l \tag{2.18}$$

nd evaluate $f(\theta_1)$ and $f(\theta_2)$. If $f(\theta_1) < f(\theta_2)$ we take our third point, θ_3, as

$$\theta_3 = \theta^* - l \tag{2.19}$$

f however $f(\theta_1) > f(\theta_2)$ we take θ_3 as

$$\theta_3 = \theta^* + 2l \tag{2.20}$$

We now evaluate $f(\theta_3)$ and use equation (2.17) to find the next approxim-tion to the minimum of f. The process may now be repeated using the new pproximation to the minimum and possibly a reduced step length. For details ee Rao (1979).

Quadratic interpolation can be used in its own right to find the minimum of function of a single parameter, but its greatest value lies in the multi-arameter case where, as we shall see in the next chapter, it is often necessary) find the minimum of $f(\boldsymbol{\theta})$ at points on the line $\boldsymbol{\theta}_0 + \lambda\mathbf{d}$ where $\boldsymbol{\theta}_0$ is a given oint and $\mathbf{d}$ specifies a given direction. The values of $f(\boldsymbol{\theta}_0 + \lambda\mathbf{d})$ on this line are ınctions of the single variable, λ. For example, if the minimum of the ınction

$$f(\boldsymbol{\theta}) = (\theta_1^2 - \theta_2)^2 + (1 - \theta_1)^2 \tag{2.21}$$

om the point $\boldsymbol{\theta}_0' = [-2, -2]$ along the direction $\mathbf{d}' = [1.00 \;\; 0.25]$ is desired, e set the new point $\boldsymbol{\theta}_1$ as

$$\boldsymbol{\theta}_1 = \boldsymbol{\theta}_0 + \lambda\mathbf{d} \tag{2.22}$$

and find the minimum of λ, λ^* say, which makes $f(\boldsymbol{\theta}_1)$ a minimum. Since $f(\boldsymbol{\theta}_1)$ equals

$$f(-2+\lambda, -2+0.25\lambda) = [(-2+\lambda)^2 - (-2+0.25\lambda)]^2 + [1-(-2+\lambda)] \qquad (2.23)$$

we need to find the value of λ which minimizes (2.23). Quadratic interpolation may be used to find λ^*, and examples of this use of the technique will be given in Chapter 5.

A procedure similar to quadratic interpolation but one which is more accurate and efficient is that described by Davidon (1959), which involves approximating the function by a cubic polynomial. It is very widely used in the context of linear searches in the multiparameter case as we shall see in later chapters. A detailed description of the method is given in Walsh (1975).

2.3 MULTIPARAMETER SEARCH METHODS

Many direct search methods have been proposed for locating the minimum of a function of more than a single parameter, but in this section we shall consider only one, namely the *simplex* method originally described by Spendley, Hext and Himsworth (1962) and further developed by Nelder and Mead (1965). A simplex is the geometric figure formed by a set of $m+1$ points in m-dimensional space. When the points are equidistant the simplex is said to be regular; in two dimensions the simplex is a triangle and in three dimensions it is a tetrahedron.

The basic idea of the simplex method is to compare the values of the objective function at the $m+1$ vertices of a general simplex and move this simplex gradually towards the minimum during the iterative process. The original technique as proposed by Spendley *et al.* (1962) maintained a regular simplex at each stage, but Nelder and Mead (1965) proposed several modifications including allowing the simplex to become non-regular, which increased the power and the efficiency of the method. Movement of the simplex towards the minimum is achieved by using three basic operations *reflection*, *contraction* and *expansion*, each of which is described in detail below. We begin with $(m+1)$ points, $\boldsymbol{\theta}_1, \boldsymbol{\theta}_2, \ldots, \boldsymbol{\theta}_{m+1}$ and evaluate the function at each to obtain $f_i = f(\boldsymbol{\theta}_i)$ $i = 1, \ldots, m+1$. Suppose the highest of these values is f_n corresponding to the vertex, $\boldsymbol{\theta}_n$. Clearly we should move away from $\boldsymbol{\theta}_n$, and this we do by reflecting $\boldsymbol{\theta}_n$ in the opposite face of the simplex to obtain a point $\boldsymbol{\theta}_r$, say, which is used to construct a new simplex. Mathematically the reflected point is given by

$$\boldsymbol{\theta}_r = (1+\alpha)\boldsymbol{\theta}_0 - \alpha\boldsymbol{\theta}_n \qquad (2.24)$$

where $\boldsymbol{\theta}_0$ is the centroid of all the points *with the exception of* $\boldsymbol{\theta}_n$, that is

$$\boldsymbol{\theta}_0 = \frac{1}{m}\sum_{\substack{i=1 \\ i \neq n}}^{m+1} \boldsymbol{\theta}_i \tag{2.25}$$

and α is the *reflection coefficient* defined as

$$\alpha = \frac{\text{distance between } \boldsymbol{\theta}_r \text{ and } \boldsymbol{\theta}_0}{\text{distance between } \boldsymbol{\theta}_n \text{ and } \boldsymbol{\theta}_0} \tag{2.26}$$

The reflected point $\boldsymbol{\theta}_r$ will be on the line joining $\boldsymbol{\theta}_n$ and $\boldsymbol{\theta}_0$ on the far side of $\boldsymbol{\theta}_0$ from $\boldsymbol{\theta}_n$, as is illustrated in Fig. 2.2. The points $\boldsymbol{\theta}_1$, $\boldsymbol{\theta}_2$ and $\boldsymbol{\theta}_3$ form the original simplex and the points, $\boldsymbol{\theta}_1$, $\boldsymbol{\theta}_2$ and $\boldsymbol{\theta}_4$ form the new one.

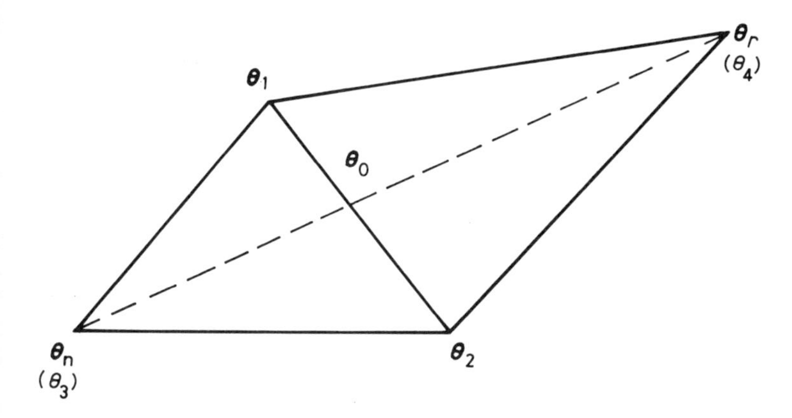

Figure 2.2 Reflection step of the simplex minimization technique.

Since the direction of movement of the simplex is always away from the highest function value, repetitive application of the reflection process will, in general, lead to a zig-zag path in the direction of the minimum, as shown in Fig. 2.3.

If only the reflection process is used for finding the minimum, difficulties may be encountered when, for example, the simplex straddles a 'valley' and the values of the objective function for $\boldsymbol{\theta}_n$ and $\boldsymbol{\theta}_r$ are equal, since a closed cycle of operations will result. This situation is illustrated in Fig. 2.4. Here $\boldsymbol{\theta}_2$ is the worst point in the simplex defined by the vertices $\boldsymbol{\theta}_1$, $\boldsymbol{\theta}_2$ and $\boldsymbol{\theta}_3$ and the reflection procedure leads to the new simplex, $\boldsymbol{\theta}_1$, $\boldsymbol{\theta}_2$, $\boldsymbol{\theta}_4$. Now $\boldsymbol{\theta}_4$ has the highest function value and so reflection leads back to the original simplex. Consequently the search for the minimum is stranded over the valley. This problem is overcome by ensuring that no return can be made to a point which has just been left, using the process of expansion or contraction.

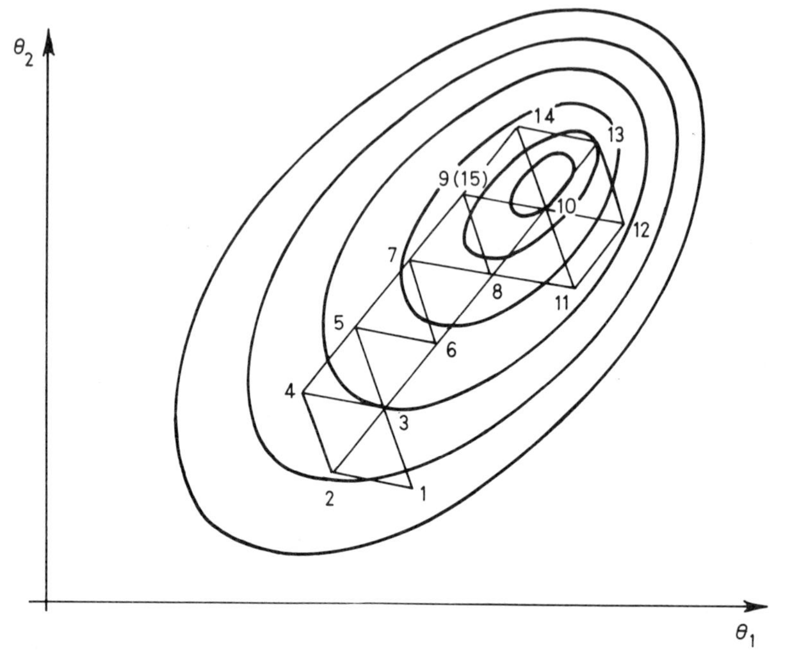

Figure 2.3 Simplex search.

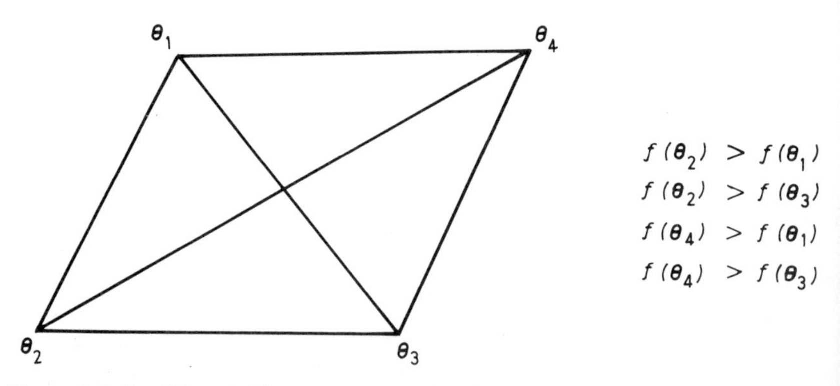

Figure 2.4 Possible problem encountered with simplex procedure when using reflection step.

For example, if the reflection procedure leads to a point $\boldsymbol{\theta}_r$ for which $f(\boldsymbol{\theta}_r) > f(\boldsymbol{\theta}_i)$ for all i except $i = n$, then the simplex is contracted along the direction $\boldsymbol{\theta}_r$ to $\boldsymbol{\theta}_0$ in the hope of finding a better point from which to return to the reflection procedure. Alternatively if the reflection produces a new minimum, it might be that the function value could be further decreased by continuing along the direction $\boldsymbol{\theta}_0$ to $\boldsymbol{\theta}_r$, in other words by expanding the simplex.

The convergence criterion generally employed with the simplex method is that the standard deviation of the function values at the $m+1$ vertices of the current simplex is less than small value, ϵ. That is

$$\left\{\frac{\sum_{i=1}^{m+1}[f(\boldsymbol{\theta}_i)-f(\boldsymbol{\theta}_0)]^2}{m+1}\right\}^{1/2} < \epsilon \tag{2.27}$$

To illustrate the use of the simplex method we shall apply it to finding the minimum of the function

$$f(\theta_1, \theta_2) = 100(\theta_1 - \theta_2)^2 + (1-\theta_1)^2 \tag{2.28}$$

starting from the point $\theta_1 = 0$, $\theta_2 = 0$. A contour plot of this function is shown in Fig. 2.5; it was first suggested as a test problem for minimization algorithms by Rosenbrock (1960). The steps taken by the simplex method during its first few iterations are shown in Table 2.2. With $\epsilon = 0.0001$ the technique converged in 63 iterations to the known minimum of $\theta_1 = \theta_2 = 1$.

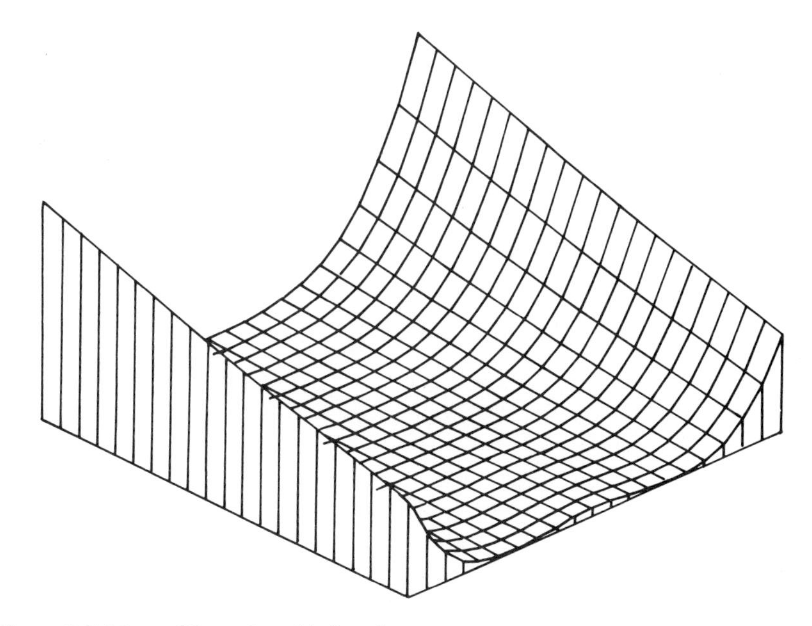

Figure 2.5 View of Rosenbrock's function.

Table 2.2 Simplex method applied to Rosenbrock's test function

	Coordinates of vertices of simplex		
Iteration	1	2	3
1	[0.00, 0.00]	[0.43, 0.00]	[0.00, 0.10]
2	[0.00, 0.00]	[−0.21, 0.07]	[0.00, 0.10]
3	[0.00, 0.00]	[−0.21, 0.07]	[−0.16, 0.01]
10	[0.21, 0.01]	[0.16, 0.02]	[0.11, 0.01]
20	[0.43, 0.19]	[0.44, 0.18]	[0.49, 0.23]
40	[0.97, 0.94]	[0.97, 0.93]	[0.95, 0.91]

2.4 SUMMARY

Direct search methods minimize a function by a series of function evaluations at particular points. The interpolation method discussed in Section 2.2 is important not only for the minimization of functions of a single parameter but also for linear searches in the multiparameter case. Statistical examples of the techniques described will be given in later chapters.

3

Gradient methods

3.1 INTRODUCTION

In this chapter we shall consider a number of minimization methods which require the evaluation of the derivatives of the function as well as the function values themselves. Much of the theory surrounding such methods is strictly applicable only to *quadratic functions* (see Section 3.3), but fortunately many objective functions can be well approximated by quadratics in the neighbourhood of the minimum. We begin with an account of the simplest and possibly oldest of this type of minimization procedure, steepest descent.

3.2 THE METHOD OF STEEPEST DESCENT

The partial derivatives of a function, $f(\boldsymbol{\theta})$, with respect to each of the m parameters in $\boldsymbol{\theta}$ are collectively called the *gradient* of the function. The gradient vector, $\mathbf{g}$, is then simply given by

$$\mathbf{g}' = \frac{\partial f}{\partial \theta_1}, \frac{\partial f}{\partial \theta_2}, \ldots, \frac{\partial f}{\partial \theta_m} \tag{3.1}$$

The gradient vector has a very important property. If we move in a direction defined by $\mathbf{g}$ from any point in m-dimensional space, the function value increases at the fastest possible rate, and if we move in the direction of the negative gradient, the function value decreases at the fastest possible rate. (For a proof of this property see Box, Davies and Swann, 1969.)

The method of steepest descent seeks to exploit this property of the gradient direction. Given the current point $\boldsymbol{\theta}_i$, the point $\boldsymbol{\theta}_{i+1}$ is obtained by a linear search (see Chapter 2) in the direction $-\mathbf{g}(\boldsymbol{\theta}_i)$ (i.e. the gradient vector evaluated at the current point $\boldsymbol{\theta}_i$). The iterative process to find the minimum is thus

$$\boldsymbol{\theta}_{i+1} = \boldsymbol{\theta}_i - \lambda_i \mathbf{g}(\boldsymbol{\theta}_i) \tag{3.2}$$

where the initial point $\boldsymbol{\theta}_1$ is given and λ_i is determined by a linear search procedure.

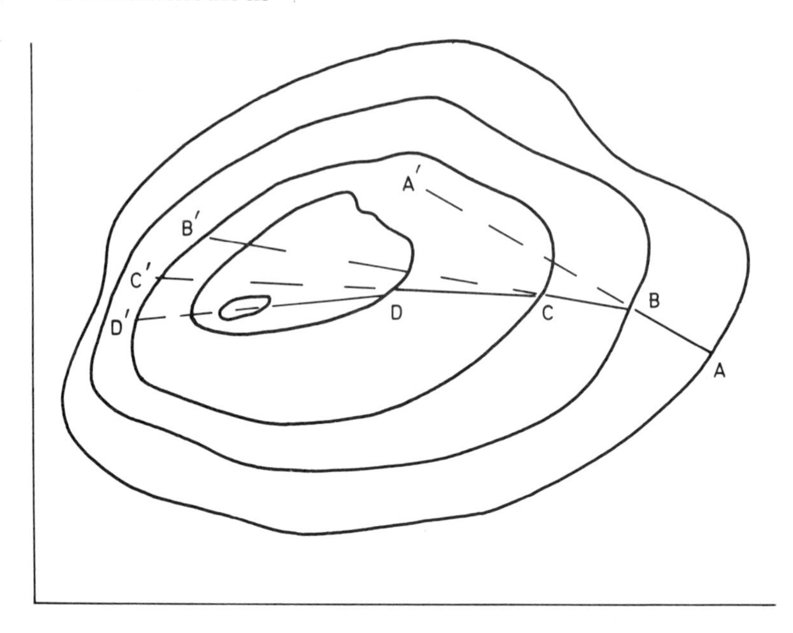

Figure 3.1 Steepest descent minimization.

The method of steepest descent might on first consideration appear to be the best method for minimizing a function since each one-dimensional search begins in the 'best' direction. Unfortunately, however, the direction of steepest descent is a local rather than a global property and so frequent changes of direction are often necessary, making the method very inefficient for many problems. The problem is illustrated in Fig. 3.1 where the negative gradient vectors evaluated at points A, B, C and D are along the directions AA′, BB′, CC′, DD′. Thus the function value decreases at the fastest rate in direction AA′ at point A but *not* at point B. Similarly the function value decreases at the fastest rate in the direction BB′ at point B but not at point C, and so on.

Convergence of the steepest descent method can be painfully slow and numerous modifications have been suggested over the years to improve its speed. For example, a modification proposed by Forsythe and Motzkin (1951) is to use the search direction given by

$$\boldsymbol{\theta}_i - \boldsymbol{\theta}_{i-2} \tag{3.3}$$

from time to time in place of the negative gradient. This modification generally results in faster convergence than by straightforward steepest descent.

To illustrate the use of steepest descent we shall apply the method

Table 3.1 Steepest descent on Rosenbrock's test function

Iteration	θ_1	θ_2
1	0.16	0.00
2	0.16	0.03
10	0.29	0.07
50	0.48	0.23
100	0.58	0.34
500	0.83	0.68

to Rosenbrock's test function (see equation (2.28)); starting values of $\theta_1 = \theta_2 = 0.0$ are used and the results are summarized in Table 3.1. The procedure is extremely slow to converge for this particular function; for example, if as convergence criterion we require that the length of the gradient vector should be less than 0.001, then over 4000 iterations are needed and the estimated minimum is $\theta_1 = 0.999$, and $\theta_2 = 0.998$.

A major problem with the method of steepest descent is that no account is taken of the second derivatives of $f(\boldsymbol{\theta})$ and yet the curvature of the function – which determines its behaviour near the minimum – depends on these derivatives. The Newton–Raphson method, described in the following section, overcomes this disadvantage.

3.3 THE NEWTON–RAPHSON METHOD

A quadratic function is defined as follows:

$$f(\boldsymbol{\theta}) = a + \boldsymbol{\theta}'\mathbf{b} + \tfrac{1}{2}\boldsymbol{\theta}'\mathbf{B}\boldsymbol{\theta} \tag{3.4}$$

where a is a scalar, $\mathbf{b}$ is a vector whose elements are constants and $\mathbf{B}$ is a positive definite symmetric matrix. Such a function has its minimum at a point $\boldsymbol{\theta}^*$ given by

$$\boldsymbol{\theta}^* = -\mathbf{B}^{-1}\mathbf{b} \tag{3.5}$$

(found by simply equating the derivatives of f with respect to $\boldsymbol{\theta}$ to the null vector).

By use of Taylor's expansion it is easy to show that, subject to certain continuity conditions, a function $f(\boldsymbol{\theta})$ may be approximated in the region of a point $\boldsymbol{\theta}_0$ by a function $h(\boldsymbol{\theta})$ given by

$$h(\boldsymbol{\theta}) = f(\boldsymbol{\theta}_0) + (\boldsymbol{\theta} - \boldsymbol{\theta}_0)'\mathbf{g}(\boldsymbol{\theta}_0) + \tfrac{1}{2}(\boldsymbol{\theta} - \boldsymbol{\theta}_0)'\mathbf{G}(\boldsymbol{\theta}_0)(\boldsymbol{\theta} - \boldsymbol{\theta}_0) \tag{3.6}$$

where $\mathbf{g}$ is the gradient vector described in the previous section and $\mathbf{G}$ is an

$(m \times m)$ matrix of second derivatives of f, that is the Hessian matrix defined by equation (1.8) of Chapter 1.

A reasonable approximation to the minimum of $f(\boldsymbol{\theta})$ might be the minimum of $h(\boldsymbol{\theta})$, $\boldsymbol{\theta}^*$, which by comparing terms in (3.6), (3.5) and (3.4) is easily found to be

$$\boldsymbol{\theta}^* = \boldsymbol{\theta}_0 - \mathbf{G}^{-1}(\boldsymbol{\theta}_0)\mathbf{g}(\boldsymbol{\theta}_0) \tag{3.7}$$

The Newton–Raphson method uses $\boldsymbol{\theta}^*$ as the next approximation to the minimum giving therefore the iterative scheme

$$\boldsymbol{\theta}_{i+1} = \boldsymbol{\theta}_i - \mathbf{G}^{-1}(\boldsymbol{\theta}_i)\mathbf{g}(\boldsymbol{\theta}_i) \tag{3.8}$$

or more generally

$$\boldsymbol{\theta}_{i+1} = \boldsymbol{\theta}_i - \lambda_i\mathbf{G}^{-1}(\boldsymbol{\theta}_i)\mathbf{g}(\boldsymbol{\theta}_i) \tag{3.9}$$

where λ_i is determined by a linear search from $\boldsymbol{\theta}_i$ in the direction $-\mathbf{G}^{-1}\mathbf{g}$.

The convergence of the Newton–Raphson method is very fast when $\boldsymbol{\theta}$ is close to the minimum, since, in general, $h(\boldsymbol{\theta})$ will provide a good approximation to $f(\boldsymbol{\theta})$ in this region. The disadvantages of the method are firstly that it calls for the evaluation and inversion of the Hessian matrix at each step and this may be a major computational burden if the number of parameters is at all large, and secondly that if $\boldsymbol{\theta}_i$ is not close to the minimum, $\mathbf{G}$ may become negative definite, in which case the method may fail to converge.

When the Newton–Raphson method is applied to Rosenbrock's test function using starting values of $\theta_1 = 0$ and $\theta_2 = 0$, the minimum at $\theta_1 = 1.0$, $\theta_2 = 1.0$ is found after nine iterations (when the same convergence criterion as that for the steepest descent example is used).

Comparison of equations (3.2) and (3.9) shows that in steepest descent the direction of search is $-\mathbf{g}(\boldsymbol{\theta}_i)$ and in Newton–Raphson it is $-\mathbf{G}^{-1}(\boldsymbol{\theta}_i)\mathbf{g}(\boldsymbol{\theta}_i)$; when $\mathbf{G} = \mathbf{I}$, the identity matrix, the two methods are identical. Several other methods attempt to find a search direction of the form $\mathbf{H}(\boldsymbol{\theta}_i)\mathbf{g}(\boldsymbol{\theta}_i)$ where $\mathbf{H}$ is a positive definite symmetric matrix updated at each iteration, and which eventually converges to $\mathbf{G}^{-1}$. The resulting iterative scheme has the advantage of avoiding the inversion of an $(m \times m)$ matrix at each stage and by ensuring that $\mathbf{H}$ remains positive definite also avoids the second problem of the Newton–Raphson method mentioned above. Such methods are collectively known as quasi-Newton procedures and differ in the way in which $\mathbf{H}$ is chosen.

3.4 THE DAVIDON–FLETCHER–POWELL METHOD

One of the most powerful of the class of *quasi-Newton* minimization procedures is that developed by Davidon (1959) and Fletcher and Powell

1963). Essentially this method begins as steepest descent and changes over to Newton's method during the course of a number of iterations, by continually pdating an approximation to the inverse of the matrix of second derivatives t the minimum in such a way as to ensure that the matrices $\{\mathbf{H}_i\}$ are positive efinite. The basic iteration of this method is

$$\boldsymbol{\theta}_{i+1} = \boldsymbol{\theta}_i - \lambda_i \mathbf{H}_i \mathbf{g}_i \tag{3.10}$$

here $\mathbf{g}_i$ is the gradient vector evaluated at $\boldsymbol{\theta}_i$, and $\mathbf{H}_i$ is the ith approximation the inverse of the matrix of second derivatives. The step λ_i to the minimum long the search direction may be found using one of the univariate search rocedures described in the previous chapter. The initial choice of $\mathbf{H}$ is rbitrary provided it is positive definite, and the identity matrix is usually hosen with the consequence that the first step is taken in the direction of eepest descent.

The approximation to the inverse matrix of second derivatives is updated sing the relation

$$\mathbf{H}_{i+1} = \mathbf{H}_i + \mathbf{A}_i + \mathbf{B}_i \tag{3.11}$$

here

$$\mathbf{A}_i = \mathbf{z}\mathbf{z}'/\mathbf{z}'\mathbf{u} \tag{3.12}$$

$$\mathbf{B}_i = \mathbf{H}_i\mathbf{u}\mathbf{u}'\mathbf{H}_i/\mathbf{u}'\mathbf{H}_i\mathbf{u} \tag{3.13}$$

nd

$$\mathbf{z} = -\lambda_i \mathbf{H}_i \mathbf{g}_i \tag{3.14}$$

$$\mathbf{u} = \mathbf{g}_{i+1} - \mathbf{g}_i \tag{3.15}$$

The justification for this procedure is given in Fletcher and Powell (1963). ssentially the matrix $\mathbf{A}_i$ ensures that the sequence of matrix approximations onverges to $\mathbf{G}^{-1}$, and $\mathbf{B}_i$ ensures that each estimate $\mathbf{H}_i$ is positive definite. etcher and Powell show that for a quadratic function $\mathbf{H}$ will be equal to $\mathbf{G}^{-1}$ ter m iterations.

The results of applying this method to Rosenbrook's function are shown in able 3.2.

5 THE FLETCHER–REEVES METHOD

early a critical factor in the efficiency of any interactive minimization ocedure is the direction of search at each stage. An important concept here that of *conjugate* directions, since for a quadratic function the 'best' search rection is in a direction conjugate to that taken on the previous step, where

Table 3.2 Quasi–Newton procedure applied to Rosenbrock's test function

Iteration	θ_1	θ_2
1	0.16	0.00
2	0.29	0.05
3	0.34	0.12
4	0.45	0.19
5	0.53	0.26
10	1.02	1.05
15	1.00	1.00

two directions **p** and **q** are said to be conjugate with respect to the symmetr positive definite matrix **G** if

$$\mathbf{p}'\mathbf{G}\mathbf{q} = 0 \quad (3.1$$

It is easy to show (see, for example, Walsh, 1975) that if searches are carri out in mutually conjugate directions then the minimum of a quadra function of m parameters will be found in at most m steps. The Fletche Reeves method attempts to exploit this fact and uses a simple recurren formula which produces a sequence of mutually conjugate directions. T initial step is in the direction of steepest descent and subsequent directions a found from

$$\mathbf{d}_i = -\mathbf{g}_i + (||\mathbf{g}_i||^2/||\mathbf{g}_{i-1}||^2)\mathbf{d}_{i-1} \quad (3.1$$

These directions are mutually conjugate and the method will find t minimum of a quadratic function of m variables after at most m steps. Wh

Table 3.3 Fletcher–Reeves method on Rosenbrock test function

Iteration	θ_1	θ_2
1	0.16	0.00
2	0.29	0.04
3	0.50	0.19
4	0.47	0.22
5	0.67	0.42
6	0.67	0.42
7	0.66	0.43
8	0.79	0.62
9	0.84	0.68
10	0.83	0.69

pplied to non-quadratic functions it will hope to achieve the quadratic onvergence property when the quadratic approximation becomes valid. letcher and Reeves suggest that every mth search direction should be along ıe direction of steepest descent and that the construction of conjugate irections should then restart.

Applying the Fletcher–Reeves method to Rosenbrock's function gives the esults shown in Table 3.3.

.6 SUMMARY

he minimization methods described in this chapter involve the evaluation of ıe derivatives of the objective function as well as the function values themelves in seeking the minimum. The simplest technique, steepest descent, ay in practice only converge extremely slowly and is not recommended. The ewton–Raphson method involves the inversion of an $(m \times m)$ matrix at each ep, which computationally may be very expensive; it may also *diverge* if the itial approximation to the minimum is some distance away from the true lue. The Davidon–Fletcher–Powell and Fletcher–Reeves methods overome such problems in many situations and have become perhaps the most nportant minimization methods in practice, particularly with regard to atistical problems, as we shall see in later chapters.

4

Some examples of the application of optimization techniques to statistical problems

4.1 INTRODUCTION

In this chapter we shall illustrate the use of several of the minimizati
techniques discussed in Chapters 2 and 3 on a number of relatively straig
forward statistical problems. In addition a number of optimization algorith
not previously described but which are of particular relevance in statistics w
be mentioned.

4.2 MAXIMUM LIKELIHOOD ESTIMATION

The maximum likelihood principle and its use in estimation is extrem
important in statistics and is discussed in detail in Kendall and Stuart volun
1 and 2 (1979). Essentially maximum likelihood estimation involves es
mating a parameter (or several parameters) by maximizing the *likeliho*
function, this being simply the joint probability of the observations regard
as a function of the parameter(s). So given a sample of n values, $x_1, \ldots,$
from the density function $f(x;\theta)$ dependent on the single parameter, θ, t
likelihood function is given by

$$\mathscr{L}(x_1, \ldots, x_n; \theta) = f(x_1; \theta) f(x_2; \theta) \ldots f(x_n; \theta) \qquad (4$$

The maximum likelihood principle implies that we take as our estimator c
that value, $\hat{\theta}$ say, within the admissible range of θ which makes the likelih
function as large as possible.

If the likelihood function is a twice differentiable function of θ then $\hat{\theta}$ wil
given by the root of

$$\frac{d\mathscr{L}}{d\theta} = 0, \qquad (4$$

›r which

$$\frac{d^2\mathscr{L}}{d\theta^2} < 0 \tag{4.3}$$

If (4.2) and (4.3) are satisfied $\hat{\theta}$ will correspond to a possibly *local* maximum f the likelihood function. The value of $\hat{\theta}$ corresponding to the largest of these f there are in fact more than one) will be the required estimate of θ. (In ractice it is often simpler to work with the logarithm of the likelihood nction which, since it has its maxima at the same positions as $\mathscr{L}$, will lead to e same estimated value for θ.)

Maximum likelihood estimators have a number of desirable statistical operties such as consistency, asymptotic normality, etc., which are scribed in detail in Kendall and Stuart (1979). Here, however, we shall ncentrate on methods for finding the estimators. In some cases equation .2) will have an explicit solution giving the estimator as some function of the servations. An example of this situation was given in Chapter 1. More teresting for our purpose, however, are those problems where (4.2) cannot solved explicitly and where we must turn to one or other of the algorithms scribed in the previous two chapters to find our estimate of θ.

We shall begin with a relatively straightforward example, which involves *t* the estimation of the parameter of a simple Poisson distribution but of a *ncated* Poisson of the following form

$$f(x;\theta) = \frac{e^{-\theta}\theta^x}{x!(1-e^{-\theta})} \qquad x = 1, 2, \ldots \tag{4.4}$$

uch a density might arise, for example, in studying the size of groups at rties.) The log-likelihood function in this case is

$$L = -n\theta + \log_e\theta \sum_{i=1}^{n} x_i - \sum_{i=1}^{n} \log_e x_i! - n\log_e(1-e^{-\theta}) \tag{4.5}$$

$$\frac{dL}{d\theta} = -n + \frac{\Sigma x_i}{\theta} - \frac{ne^{-\theta}}{1-e^{-\theta}}. \tag{4.6}$$

we set $(dL/d\theta)$ to zero we find that the resulting equation has no explicit ution for θ; consequently we are led to consider one or other of the merical techniques discussed in Chapters 2 and 3. In the context of a atively simple estimation situation such as this, where it is straightforward obtain first and second derivatives of L, perhaps the most obvious method to use is Newton–Raphson (see Chapter 3); the second derivative of L wit respect to θ is

$$\frac{d^2L}{d\theta^2} = -\frac{\sum_{i=1}^{n} x_i}{\theta^2} + \frac{ne^{-\theta}}{(1-e^{-\theta})^2} \quad (4.$$

and using (4.6) and (4.7) we shall apply the method to the problem estimating the parameter of a truncated Poisson distribution using the da shown in Table 4.1. (Here the observations are grouped but since this involv only a trivial change to the estimation procedure, the details are left to th reader as an exercise.)

As a starting value for the iterative procedure we shall use the mean valu of the observations which takes the value 1.5118. The steps taken by th Newton–Raphson method are shown in Table 4.2; the final estimate of obtained is the value 0.8925. (The convergence criterion used was that th difference between successive parameter estimates was less than 0.001.)

Table 4.1 Grouped data from a truncated Poisson distribution

x	f_x
1	1486
2	694
3	195
4	37
5	10
6	1

Table 4.2 Newton–Raphson applied to finding the m.l.e. of the parameter ir truncated Poisson density, using the data in Table 4.1

Iteration	$\frac{\partial L}{\partial \theta}$	$\frac{\partial^2 L}{\partial \theta^2}$	θ	L
1	−685.5137	−723.3218	1.5118	−1545.55
2	873.5129	−4095.9727	0.5641	−1425.14
3	228.3008	−2248.1108	0.7773	−1314.77
4	24.1609	−1799.6599	0.8789	−1302.34
5	0.3252	−1751.4932	0.8923	−1302.17
6	0.0024	−1750.8418	0.8925	−1302.17

The same method was applied to the data of Table 4.1 using a number of ther starting values for θ. For example, a starting value of 1.8 again led to a nal value of 0.8925, although two more iterations were necessary to achieve ne same convergence criterion. When starting values of 2.0 or above were sed the procedure *diverged*, illustrating the problem of the Newton–aphson method mentioned in Chapter 3.

A modification of the Newton–Raphson method much used in statistical pplications is the so-called *Fisher's method of scoring*. This method has as its asic step the following:

$$\theta_{i+1} = \theta_i - \left[E\left(\frac{d^2L}{d\theta^2}\right)\right]^{-1}_{\theta_i}\left[\frac{dL}{d\theta}\right]_{\theta_i} \tag{4.8}$$

nat is $d^2L/d\theta^2$ in the Newton–Raphson method is replaced by its expected alue. For the truncated Poisson example we need to find the expected value f the expression given in equation (4.7). Essentially this involves simply nding the expected value of the truncated Poisson distribution; from (4.4) nis is defined to be

$$E(x) = \sum_{x=1}^{\infty} xe^{-\theta}\theta^x/x!(1-e^{-\theta}) \tag{4.9}$$

$$= \theta/(1-e^{-\theta}) \tag{4.10}$$

sing this result to implement Fisher's scoring method and applying it to the ata of Table 4.1 with a starting value of 1.5118 gives the results shown in able 4.3. Convergence takes place a little faster than for Newton–Raphson, nd the same final value of θ is obtained.

More interesting is the difference between the two methods when poor arting values are used. For example, Table 4.4 shows the results from the oring algorithm using a starting value for θ of 6.0. As mentioned previously, e Newton–Raphson procedure will not converge from such a starting value.

ble 4.3 Fisher's method of scoring applied to finding the m.l.e. of θ in a truncated oisson density using the data in Table 4.1

ration	$\frac{\partial L}{\partial \theta}$	$\frac{\partial^2 L}{\partial \theta^2}$	θ	L
	−685.5137	−1176.7632	1.5118	−1545.5549
	−62.0889	−1696.2834	0.9293	−1303.3340
	−0.2822	−1750.5906	0.8927	−1302.1790
	0.0012	−1750.8389	0.8925	−1302.1792

Table 4.4 Fisher's method of scoring applied to finding the m.l.e. of θ in a truncate Poisson density using the data in Table 4.1

Iteration	$\frac{\partial L}{\partial \theta}$	$\frac{\partial^2 L}{\partial \theta^2}$	θ	L
1	−1818.5208	−398.8008	6.0000	−7968.7695
2	−631.5933	−1219.1560	1.4400	−1498.2639
3	−50.1289	−1706.7615	0.9220	−1302.9250
4	−0.1809	−1750.6760	0.8926	−1302.1794
5	0.0010	−1750.8359	0.8925	−1302.1792

Table 4.5
(a) Quadratic interpolation [0.1, 3.0]

Iteration	θ	L
1	1.2600	−1398.1243
2	1.8400	−1805.6992
3	0.6800	−1348.5460
4	0.8897	−1302.1862
5	0.9184	−1302.7556
6	0.8942	−1302.1818
7	0.8925	−1302.1792

(b) Cubic interpolation [0.1, 3.0]

Iteration	θ	L
1	1.5500	−1572.2624
2	0.8250	−1306.3578
3	0.8993	−1302.1792

For a single-parameter problem such as this we might also consider usi one of the search methods described in Chapter 2. The results of applying tv of these methods, quadratic and cubic interpolation, to search in the interv (0.1, 3.0) are shown in Table 4.5. The superiority of the cubic interpolati method is clearly demonstrated.

Maximum likelihood is also often used for the simultaneous estimation several parameters, particularly in the area of multivariate statistics as shall see in the next chapter. As a simple example we shall consider the estimation of the three parameters of the following gamma distribution:

$$f(x;\alpha,\sigma,p) = \frac{1}{\sigma\Gamma(p)}\left(\frac{x-\alpha}{\sigma}\right)^{p-1} \exp\left\{-\left(\frac{x-\alpha}{\sigma}\right)\right\} \tag{4.11}$$

$$\alpha \leq x \leq \infty, \sigma > 0, p > 2.$$

The log-likelihood function is given by

$$L = -np\log\sigma - n\log\Gamma(p) + (p-1)\sum_{i=1}^{n}\log(x_i-\alpha) - \sum_{i=1}^{n}(x_i-\alpha)/\sigma \tag{4.12}$$

The likelihood equations are derived by differentiating L with respect to α, σ and p and setting each derivative to zero:

$$\frac{\partial L}{\partial \alpha} = -(p-1)\sum_{i=1}^{n}(x_i-\alpha)^{-1} + n/\sigma = 0, \tag{4.13}$$

$$\frac{\partial L}{\partial \sigma} = -np/\sigma + \sum_{i=1}^{n}(x_i-\alpha)/\sigma^2 = 0, \tag{4.14}$$

$$\frac{\partial L}{\partial p} = -n\log\sigma - n\frac{\mathrm{d}}{\mathrm{d}p}\log\Gamma(p) + \sum_{i=1}^{n}\log(x_i-\alpha) = 0 \tag{4.15}$$

These three simultaneous equations cannot be solved explicitly for the three parameters and so some type of numerical method must be employed. An added problem here is that the parameters are subject to constraints; these are, however, relatively straightforward and can be dealt with by a reparameterization of the form

$$\begin{aligned} \sigma &= u^2 \\ p &= 2 + v^2 \\ \alpha &= x_{\min} - w^2 \end{aligned} \tag{4.16}$$

where $x_{\min}$ is the minimum value in the sample. An unconstrained minimization may now be performed in u, v and w.

One hundred observations were generated from such a distribution with $\alpha = 5.0$, $\sigma = 1.0$ and $p = 6.0$; a histogram of the data is shown in Fig. 4.1. The simplex minimization procedure described in Chapter 2 was applied to the data using as initial values for the parameters $\alpha = 7.23$ (the minimum value in the sample), $\sigma = 4.0$ and $p = 3.0$, corresponding to $u = 2.0$, $v = 1.0$

Table 4.6 Simplex minimization results when estimating the parameters of a gamma distribution fitted to the data shown in Fig. 4.1

	Parameter			
	u	v	w	−log-likelihood
Initial value	1.00	2.00	0.00	
Iteration 20	0.97	1.23	1.65	213.3
Iteration 40	0.90	1.14	0.51	207.0
Iteration 60	0.91	1.17	0.52	206.7
Iteration 80	1.78	0.90	0.81	204.7
Iteration 100	2.67	0.79	1.53	203.7
Final value	2.52	0.82	1.40	203.6

These values of u, v and w correspond to the following values for α, p and σ:

$$\alpha = 5.27$$
$$\sigma = 0.67$$
$$p = 8.25$$

and $w = 0.0$. With this method the derivatives in (4.13), (4.14) and (4.15) are not used; instead a direct attack is made on minimizing the negative log-likelihood. The steps taken by the simplex procedure are summarized in Table 4.6.

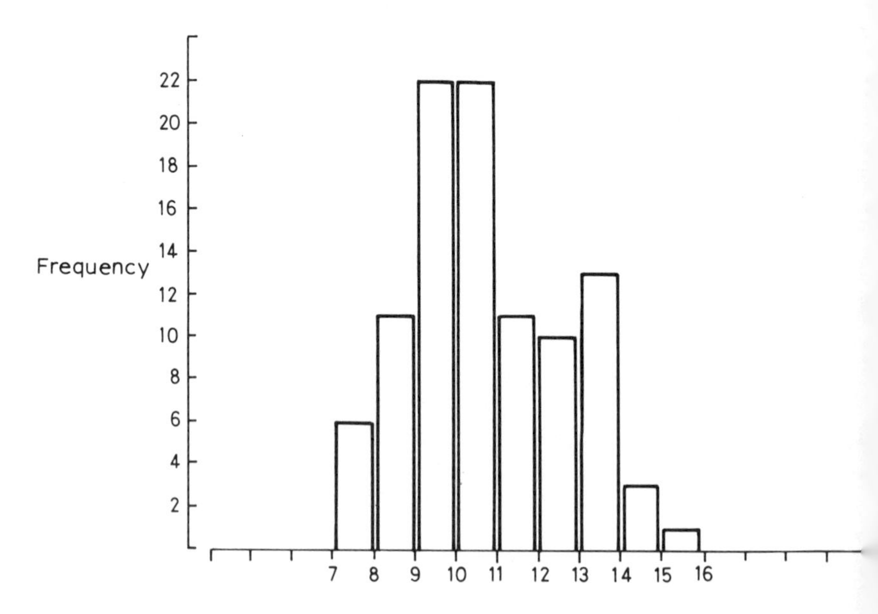

Figure 4.1 Histogram of data generated from a gamma distribution.

A variety of other initial values were tried for u, v and w, all of which led to the same final values as those given in Table 4.6. This example illustrates that with an efficient minimization procedure many maximum likelihood estimation problems might be dealt with by dealing with the likelihood function directly rather than using the derivatives which, on occasions, may be difficult to find. (Such is the case with the above example; the evaluation of the derivatives given by (4.15) is not trivial and the evaluation of second derivatives as required by Newton–Raphson, for example, would be even more difficult.)

4.3 MAXIMUM LIKELIHOOD ESTIMATION FOR INCOMPLETE DATA

Many of the most interesting estimation problems in statistics involve situations with *missing data*. A general approach to such problems is one involving maximum likelihood in association with a particular algorithm, which we shall introduce by way of a number of examples.

The first is that used by Dempster, Laird and Rubin (1977) which involves a multinomial variable $\mathbf{x} = [x_1 \; x_2 \; x_3 \; x_4 \; x_5]$. A genetic model for the population specifies cell probabilities

$$\tfrac{1}{2}, \tfrac{1}{4}\Pi, \tfrac{1}{4}(1-\Pi), \tfrac{1}{4}(1-\Pi), \tfrac{1}{4}\Pi \tag{4.17}$$

for some Π with $0 \leq \Pi \leq 1$; given a set of values for x_1, x_2, x_3, x_4 and x_5, we may find the maximum likelihood estimator of Π as follows:

$$\mathscr{L} \propto (\tfrac{1}{2})^{x_1}(\tfrac{1}{4}\Pi)^{x_2}(\tfrac{1}{4}-\tfrac{1}{4}\Pi)^{x_3}(\tfrac{1}{4}-\tfrac{1}{4}\Pi)^{x_4}(\tfrac{1}{4}\Pi)^{x_5} \tag{4.18}$$

Consequently the log-likelihood is

$$L \propto x_1 \log_e \tfrac{1}{2} + x_2 \log_e \tfrac{1}{4}\Pi + x_3 \log_e \tfrac{1}{4}(1-\Pi) + x_4 \log_e \tfrac{1}{4}(1-\Pi) + x_5 \log_e \tfrac{1}{4}\Pi \tag{4.19}$$

Differentiating L with respect to Π gives

$$\frac{\mathrm{d}L}{\mathrm{d}\Pi} = \frac{x_2+x_5}{\Pi} - \frac{x_3+x_4}{1-\Pi} \tag{4.20}$$

Setting this equal to zero leads to the following estimator for Π:

$$\hat{\Pi} = (x_2+x_5)/(x_2+x_3+x_4+x_5) \tag{4.21}$$

Suppose now that instead of x_1, x_2, x_3, x_4, x_5 the observed values consist of y_1, y_2, y_2, y_4 where $y_1 = x_1 + x_2$, $y_2 = x_3$, $y_3 = x_4$ and $y_4 = x_5$. Consequently we do not have the complete data necessary for estimating Π according to (4.21). Is it still possible to find some method of estimating Π? By using the following procedure we find that Π can indeed be estimated from the incomplete data, y_1, y_2, y_3, y_4.

1. Given a value of Π use the observed data to estimate the complete data x_1, x_2, x_3, x_4, and x_5; since x_3, x_4, and x_5 are known we need only to estimate x_1 and x_2. Clearly these will be

$$x_1 = y_1 \frac{\frac{1}{2}}{\frac{1}{2}+\frac{1}{4}\Pi}, \tag{4.22}$$

$$x_2 = y_1 \frac{\frac{1}{4}\Pi}{\frac{1}{2}+\frac{1}{4}\Pi} \tag{4.23}$$

2. Use the estimated complete data in (4.21) to obtain a revised estimate of Π.
3. Alternate steps 1 and 2 until some suitable convergence criterion is satisfied.

The results of such a procedure when the incomplete data are $y_1 = 125$, $y_2 = 18$, $y_3 = 20$ and $y_4 = 34$, and the initial value of Π is 0.5 are shown in Table 4.7.

Table 4.7 The EM* algorithm in a simple case (reproduced with permission from Dempster *et al.*, 1977)

Iteration	Π
0	0.500 000000
1	0.608 247423
2	0.624 321051
3	0.626 488879
4	0.626 777323
5	0.626 815632
6	0.626 820719
7	0.626 821395
8	0.626 821484

* See page 40.

Now let us consider a rather more complicated problem where a similar estimation procedure may be employed, namely that of estimating the five parameters in a mixture of two univariate normal distributions (a problem first considered by Karl Pearson as long ago as 1894). The relevant density function has the form

$$f(x) = pf_1(x) + (1-p)f_2(x) \tag{4.24}$$

where

$$f_i(x) = \frac{1}{\sigma_i\sqrt{2\pi}} \exp\left\{-\frac{1}{2}\left(\frac{x-\mu_i}{\sigma_i}\right)^2\right\} \qquad i = 1,2 \tag{4.25}$$

(Such density functions may be used to model a wide variety of practical problems, a number of which are discussed in Everitt and Hand, 1981.)

Given a sample of observations, $x_1, \ldots, x_n$, from the mixture density, the log-likelihood is given by

$$L = \sum_{i=1}^{n} \log_e\{pf_1(x_i) + (1-p)f_2(x_i)\} \tag{4.26}$$

Differentiating (4.26) with respect to each of the five parameters, equating the resulting expressions to zero and performing some simple algebra leads to the following series of equations:

$$\hat{p} = \frac{1}{n}\sum_{i=1}^{n} P(c_1|x_i), \tag{4.27}$$

$$\hat{\mu}_j = \frac{1}{n\hat{p}}\sum_{i=1}^{n} P(c_j|x_j)x_i \qquad j = 1,2 \tag{4.28}$$

$$\hat{\sigma}_j^2 = \frac{1}{n\hat{p}}\sum_{i=1}^{n} P(c_j|x_i)(x_i-\hat{\mu}_j)^2 \qquad j = 1,2 \tag{4.29}$$

where $P(c_j|x_i)$ is an estimate of the posterior probability that an observation x_i arises from the jth component normal density in the mixture; given a set of parameter values this estimated posterior probability would be found from

$$P(c_1|x_i) = \frac{\hat{p}f_1(x_i)}{f(x_i)}, \tag{4.30}$$

$$P(c_2|x_i) = 1.0 - P(c_1|x_i). \tag{4.31}$$

In this case the data are incomplete in the sense that we do not know which of the two component densities in the mixture each x_i value is associated with. If we did, then each posterior probability would be zero or unity and equations (4.27), (4.28) and (4.29) would reduce to estimating the mean and variance of each component using the observations known to have arisen from that component. But by an exactly analogous procedure to that described above for the simple multinomial example we can find the required estimates as follows.

1. Given values for p, μ_1, μ_2, σ_1 and σ_2, use the observed data to estimate the posterior probabilities in (4.30) and (4.31).

2. Use the estimated posterior probabilities in equations (4.27), (4.28) and (4.29) to obtain revised estimates of the five parameters.
3. Alternate steps 1 and 2 until some suitable convergence criterion is satisfied.

To illustrate the procedure for this problem, 50 observations were generated from the mixture density of (4.24) with $p = 0.4$, $\mu_1 = 0.0$, $\mu_2 = 3.0$, $\sigma_1^2 = 0.5$ and $\sigma_2^2 = 1.0$. The density function is shown in Fig. 4.2. The initial values supplied were $p = 0.2$, $\mu_1 = 1.0$, $\mu_2 = 2.0$, $\sigma_1^2 = 1.0$, $\sigma_2^2 = 0.5$, and the final values found after 80 iterations were $p = 0.42$, $\mu_1 = 0.00$, $\mu_2 = 3.00$, $\sigma_1^2 = 0.41$, $\sigma_2^2 = 0.54$. (The convergence criterion used was that the Euclidean distance between successive estimates of the five-dimensional parameter vector was less than 0.0001.) The fitted mixture is also shown in Fig. 4.2.

A comparison of this procedure, Newton–Raphson, Fletcher–Reeves and the simplex method for finding maximum likelihood estimates of the parameter in (4.24) is given in Everitt (1984). Fifty observations were taken from each of two normal mixture distributions. The relevant parameter values were:

Mixture 1: $p = 0.4, \ \mu_1 = 0.0, \ \mu_2 = 3.0, \ \sigma_1^2 = 0.5, \ \sigma_2^2 = 1.0,$

Mixture 2: $p = 0.2, \ \mu_1 = 0.0, \ \mu_2 = 3.0, \ \sigma_1^2 = 1.0, \sigma_2^2 = 2.0.$

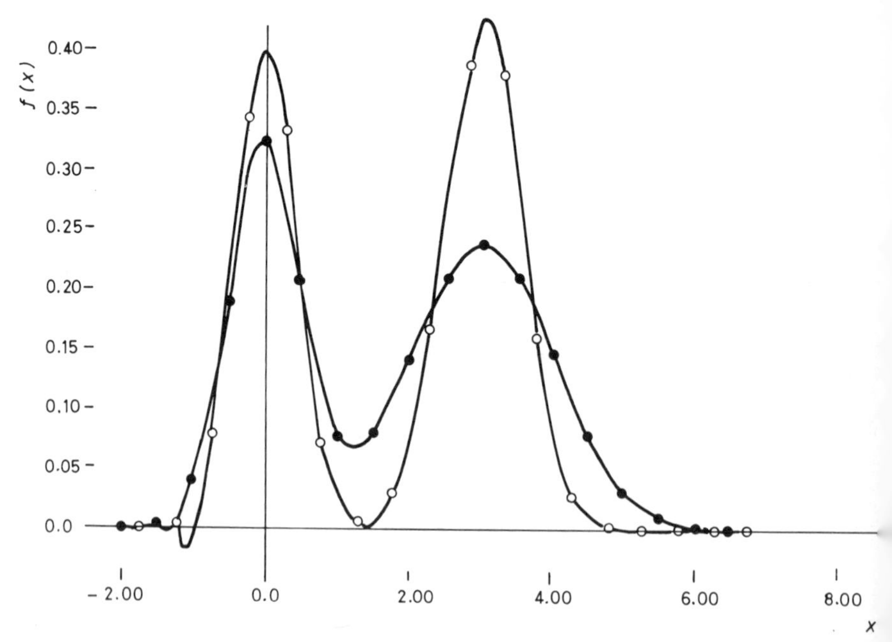

Figure 4.2 Mixture of two normal distributions. Density function •; fitted mixture ο.

For mixture 1 which has fairly clearly separated components, three different sets of starting values for the parameters were considered:

1. $p = 0.4,\ \mu_1 = 0.0,\ \mu_2 = 3.0,\ \sigma_1^2 = 0.5,\ \sigma_2^2 = 1.0,$
2. $p = 0.4,\ \mu_1 = 0.0,\ \mu_2 = 1.0,\ \sigma_1^2 = 0.5,\ \sigma_2^2 = 0.5,$
3. $p = 0.2,\ \mu_1 = 1.0,\ \mu_2 = 2.0,\ \sigma_1^2 = 1.0,\ \sigma_2^2 = 0.5.$

The first set are clearly very good initial values since they are the same as the parameter values used to generate the data. Sets 2 and 3 represent slightly worsening initial values. Table 4.8 shows a comparison of the convergence rates for the various algorithms used. (The convergence criterion used was as in the previous example.)

All the algorithms converge most rapidly when the initial values are equal to the parameter values used to generate the data. For the remaining two sets of initial values the results are not consistent; some converge more rapidly from set 2, others more rapidly from set 3. In some cases Fletcher–Reeves and the simplex method converge very slowly.

For mixture 2 the initial values used were as follows:

1. $p = 0.2,\ \mu_1 = 0.0,\ \mu_2 = 3.0,\ \sigma_1^2 = 1.0,\ \sigma_2^2 = 2.0,$
2. $p = 0.2,\ \mu_1 = 0.0,\ \mu_2 = 1.0,\ \sigma_1^2 = 1.0,\ \sigma_2^2 = 1.0,$
3. $p = 0.4,\ \mu_1 = 1.0,\ \mu_2 = 2.0,\ \sigma_1^2 = 0.5,\ \sigma_2^2 = 1.0.$

Again the first set of initial values are the same as the parameter values used to generate the data. Table 4.9 shows a comparison of the convergence rates for this case. Here the algorithm introduced above converges very slowly in some

Table 4.8 A comparison of convergence rates of algorithms finding maximum likelihood estimates of the parameters in a mixture of two normal densities

	Initial values		
Method	1	2	3
1. EM	73.3 (11–232)	81.8 (14–268)	76.3 (16–203)
2. Newton	8.9 (6–12)	16.8 (14–21)	13.6 (12–16)
3. Fletcher–Reeves	141.5 (72–240)	180.6 (84–436)	211.3 (127–438)
4. Simplex	235.2 (143–273)	239.2 (151–389)	322.8 (137–720)

Entries are the iteration average for 10 samples of size 50; figures in parentheses give the range.

Table 4.9 A comparison of convergence rates of algorithms finding maximum likelihood estimates of the parameters in a mixture of two normal densities

Method	Initial values 1	2	3
1. EM	188.6 (57–525)	174.4 (72–496)	138.8 (60–557)
2. Newton	11.1 (6–24)	18.8 (11–53)	16.9 (11–32)
3. Fletcher–Reeves	253.5 (137–437)	190.9 (108–445)	196.5 (122–291)
4. Simplex	330.2 (208–571)	366.7 (154–912)	323.4 (196–625)

Entries are the iteration average for 10 samples of size 50; figures in parentheses give the range.

cases. The results in Tables 4.8 and 4.9 indicate the superior convergence rate of the Newton–Raphson method for this particular example.

The algorithm introduced via these two examples is now generally known as the *EM algorithm* and is described in detailed theoretical terms by Dempster, Laird and Rubin (1977). The name derives from the two basic steps in the algorithm, the first of which involves estimating the complete data by finding their Expected values given the current estimates of the parameters, and the second of which involves using the estimated complete data to find parameter values which Maximize the likelihood. Dempster, Laird and Rubin (1977) explain how repeated application of the E and M steps leads ultimately to the maximum likelihood values of the parameters and Wu (1983) summarizes the properties of the algorithm, among which are the following:

1. The sequence of estimates obtained from the EM algorithm increases the likelihood and the sequence of likelihood values, if bounded above, converge to some value L^*.
2. Under certain regularity conditions, L^* will be a local maximum of the likelihood function.

Essentially then the EM algorithm will converge to the maximum likelihood estimates of the parameters (although a number of initial values may have to be tried). Such convergence may however be very slow, and some thought needs to be given to its possible advantages and disadvantages compared with other algorithms such as Newton–Raphson or Fisher scoring which might be considered as alternatives. The Newton–Raphson algorithm will clearly be superior from the point of view of convergence near a maximum since it converges quadratically. However, it does not always have the property of increasing the likelihood and may in some instances move towards a local *minimum*. Consequently the choice of starting value is likely to be far more critical for Newton–Raphson. In addition the repeated evaluation and storage of the Hessian matrix may not be feasible for many problems (see Chapter 6 for an example). The EM algorithm is in many cases extremely easy to implement on a computer and so, despite its often slow convergence, has become a very popular computational procedure in statistics.

4.4 SUMMARY

In this chapter we have illustrated the use of a number of the optimization methods described in Chapter 2 on several relatively straightforward statistical examples. In the next chapter we shall examine their use in more complex situations.

5

Optimization in regression problems

5.1 INTRODUCTION

Many optimization techniques described in the previous chapters deman large amounts of arithmetic, and their routine application to many statistic problems has only become possible with the advent of more powerf computers. For example, it is now over 40 years since Lawley formulated th maximum likelihood approach to factor analysis (Lawley, 1940), but its us has only become routine since the late 1960s when Karl Jöreskog too advantage of the development of more efficient optimization algorithn allied with the availability of the electronic computer, to produce a practic method of estimation. This, and other examples where the combination computer power and improvements in optimization methods have mac particular statistical techniques practical possibilities, will be the subject the next two chapters. In this chapter we shall concentrate on developments regression methods, and in Chapter 6 methods of multivariate analysis will t considered.

5.2 REGRESSION

Many problems encountered by the experimental scientist can be formulat in terms of determining the values of the parameters in a regression functi of the form

$$y = h(x_1, \ldots, x_p; \theta_1, \ldots, \theta_m) \qquad (5.$$

where y is a response variable, $x_1, \ldots, x_p$ explanatory variables a $\theta_1, \ldots, \theta_m$ the parameters to be estimated. When y is a continuous variat and the regression function, h, is linear in the parameters, the estimati problem is relatively straightforward using a *least squares* approach; the fitti function Q, given by

$$Q = \sum_{i=1}^{n} [f_i(y_i; \boldsymbol{\theta})]^2 \qquad (5$$

(where $f_i = y_i - h_i$ using an obvious notation), is differentiated with respect he parameters to give a series of estimation equations which are linear and which can be solved explicitly.

(In some cases the fitting function in (5.2) is slightly amended by allowing he deviation, f_i, to be weighted by a term w_i; this may be useful, for example, when the response variable does not have constant variance. The estimation procedure is then known as *weighted least squares*.)

There are, however, many interesting problems in which y is categorical ather than continuous or where h is non-linear, where we are still interested n assessing the effect of the explanatory variables on the response. Such problems require more complex estimation procedures than are involved in traightforward linear regression, as we shall see in the following sections.

.3 NON-LINEAR REGRESSION

n biochemistry the relationship between the concentration of bound (B) and ree (F) ligands at equilibrium in a receptor assay is given by the equation

$$B = \theta_1 F/(\theta_2 + F) \tag{5.3}$$

here θ_1 and θ_2 are known respectively as the *affinity* and *capacity* of the eceptor system. Here the regression function is non-linear in θ_1 and θ_2 and any ingenious methods have been suggested for rearranging (5.3) into a near form in order to use linear regression techniques to estimate θ_1 and θ_2. or example,

$$\frac{1}{B} = \frac{1}{\theta_1} + \frac{\theta_2}{\theta_1}\frac{1}{F} \tag{5.4}$$

plot of $1/B$ against $1/F$ could now provide estimates of θ_1 and θ_2 from the ope and the intercept of the line. Such methods have been reviewed by unn (1985), but they are now essentially obsolete since any of a number of umerical techniques may be applied directly to (5.3) to find least squares stimates of θ_1 and θ_2; for example, we might apply the simplex method, eepest descent or Newton–Raphson. In the case of a least squares function e steps involved in the latter two methods may be simplified somewhat by efining a matrix $\mathbf{A}$ given by

$$\mathbf{A} = \begin{bmatrix} \frac{\partial f_1}{\partial \theta_1} & \frac{\partial f_1}{\partial \theta_2} & \cdots & \frac{\partial f_1}{\partial \theta_m} \\ \frac{\partial f_2}{\partial \theta_1} & & & \\ \cdot & & & \\ \cdot & & & \\ \cdot & & & \\ \frac{\partial f_n}{\partial \theta_1} & & \cdots & \frac{\partial f_n}{\partial \theta_m} \end{bmatrix}. \tag{5.5}$$

(Since n usually exceeds m this matrix is not in general square.)

By differentiating (5.2) we find that the elements of the gradient vecto[r] necessary for the application of steepest descent are given by

$$\frac{\partial Q}{\partial \theta_k} = \sum_{i=1}^{n} 2f_i \frac{\partial f_i}{\partial \theta_k} \tag{5.6}$$

so that

$$\mathbf{g} = \begin{bmatrix} \frac{\partial Q}{\partial \theta_1} \\ \cdot \\ \cdot \\ \cdot \\ \frac{\partial Q}{\partial \theta_m} \end{bmatrix} = 2 \begin{bmatrix} \frac{\partial f_1}{\partial \theta_1} & \cdots & \frac{\partial f_n}{\partial \theta_1} \\ \cdot & & \\ \cdot & & \\ \cdot & & \\ \frac{\partial f_1}{\partial \theta_m} & \cdots & \frac{\partial f_n}{\partial \theta_m} \end{bmatrix} \begin{bmatrix} f_1 \\ \cdot \\ \cdot \\ \cdot \\ f_n \end{bmatrix} \tag{5.7}$$

that is

$$\mathbf{g} = 2\mathbf{A}'\mathbf{f} \tag{5.8}$$

where $\mathbf{f}' = [f_1 \ldots f_n]$.

Now differentiating (5.6) with respect to θ_j we find

$$\frac{\partial^2 Q}{\partial \theta_k \partial \theta_j} = 2 \sum_{i=1}^{n} \frac{\partial f_i}{\partial \theta_j} \frac{\partial f_i}{\partial \theta_k} + 2 \sum_{i=1}^{n} f_i \frac{\partial^2 f_i}{\partial \theta_j \partial \theta_k} \tag{5.9}$$

If we assume that the second term in (5.9) may be neglected we have

$$\frac{\partial^2 Q}{\partial \theta_k \partial \theta_j} \approx 2 \sum_{i=1}^{n} \frac{\partial f_i}{\partial \theta_j} \frac{\partial f_i}{\partial \theta_k} \tag{5.10}$$

These are the elements of the Hessian matrix $\mathbf{H}$, which may therefore [be] written in the form

$$\mathbf{H} \approx 2\mathbf{A}'\mathbf{A} \tag{5.11}$$

Using (5.8) and (5.11) the step in the Newton–Raphson minimizati[on] procedure (see Chapter 3) may now be written

$$\boldsymbol{\theta}_{i+1} = \boldsymbol{\theta}_i - (2\mathbf{A}'\mathbf{A})^{-1}(2\mathbf{A}'\mathbf{f}) \tag{5.12}$$

that is

$$\boldsymbol{\theta}_{i+1} = \boldsymbol{\theta}_i - (\mathbf{A}'\mathbf{A})^{-1}\mathbf{A}'\mathbf{f} \tag{5.13}$$

(In general $\mathbf{A}'\mathbf{A}$ is positive definite so the procedure should converge.)

A modification of (5.13) suggested first by Levenberg (1944) is to use the ollowing:

$$\boldsymbol{\theta}_{i+1} = \boldsymbol{\theta}_i - [\lambda \mathbf{I} + \mathbf{A}'\mathbf{A}]^{-1}\mathbf{A}'\mathbf{f} \tag{5.14}$$

vhere λ is a scalar which may be adjusted to control the sequence of iterations ınd $\mathbf{I}$ is the $(m \times m)$ identity matrix. When $\lambda \to \infty$, (5.14) becomes

$$\boldsymbol{\theta}_{i+1} = \boldsymbol{\theta}_i - (1/\lambda)\mathbf{A}'\mathbf{f} \tag{5.15}$$

'hich is essentially steepest descent, and when $\lambda \to 0$, (5.14) tends to the usual Newton–Raphson form, (5.13). Using λ to control the iterative procedure nables the method to take advantage of the reliable improvement in the bjective function given by steepest descent when still far from the minimum, nd the rapid convergence of Newton–Raphson when close to the minimum. larquardt (1963) describes a scheme for choosing λ at each iteration which as been found to be reasonably efficient, although Fletcher (1971) has ointed out possible difficulties. Techniques specifically designed for a least juares objective function can be expected to converge faster than the general iethods described in previous chapters.

To illustrate the application of optimization methods in least squares we iall use the data sets shown in Tables 5.1 and 5.2. The first of these is taken 'om Cressie and Keightley (1981) and involves hormone-receptor assay esults from a human mammary tumour at the Flinders Medical Centre. The econd data set, taken from Chatterjee and Chatterjee (1982), consists of inning times in men's running events in Olympic Games held between 1900 ıd 1976.

Considering first the data in Table 5.1, we wish to fit the model given by

Table 5.1 Concentration of free (F) and bound (B) ligand at equilibrium in a receptor assay (Cressie and Keightley, 1981)

Free	Bound
84.6	12.1
83.9	12.5
148.2	17.2
147.8	16.7
463.9	28.3
463.8	26.9
964.1	37.6
967.6	35.8
1925.0	38.5
1900.0	39.9

Table 5.2 Winning times (in seconds) in men's running events in Olympic Game (1900–76). (Taken with permission from Chatterjee and Chatterjee, 1982)

Year	100 metres	200 metres	400 metres	800 metres
1900	10.80	22.20	49.40	121.40
1904	11.00	21.60	49.20	116.00
1908	10.80	22.60	50.00	112.80
1912	10.80	21.70	48.20	111.90
1920	10.80	22.00	49.60	113.40
1924	10.60	21.60	47.60	112.40
1928	10.80	21.80	47.80	111.80
1932	10.30	21.20	46.20	109.80
1936	10.30	20.70	46.50	112.90
1948	10.30	21.10	46.20	109.20
1952	10.40	20.70	45.90	109.20
1956	10.50	20.60	46.70	107.70
1960	10.20	20.50	44.90	106.30
1964	10.00	20.30	45.10	105.10
1968	9.90	19.80	43.80	104.30
1972	10.14	20.00	44.66	105.90
1976	10.06	20.23	44.26	103.50

Table 5.3 Least squares estimation of the parameters θ_1 and θ_2 of (5.3) fitted to th data in Table 5.1

Iteration	Iterative procedure 1 (5.13)		Iterative procedure 2 (5.14)	
	θ_1	θ_2	θ_1	θ_2
Start	10.00	50.00	10.00	50.00
1	10.94	59.95	40.30	366.28
2	11.95	69.90	43.64	184.44
3	13.01	79.85	42.93	223.08
4	15.28	99.72	44.32	240.53
5	16.52	109.64	44.38	241.66
6	17.82	119.55	44.38	241.70
7	19.21	129.46		
8	20.68	139.35		
9	22.24	149.22		
10	23.91	159.09		
20	44.37	241.64		
24	44.38	241.70		

Final values $\theta_1 = 44.38$, $\theta_2 = 241.70$

5.3), first using the iterative scheme of (5.13) and then that of (5.14). In each :ase the initial values of the parameters were $\theta_1 = 10.0$, $\theta_2 = 50.0$. The esults are shown in Table 5.3.

As would be expected, the modified iterative procedure of (5.14) converges nuch more rapidly than that of (5.13); both methods produce the same final values for θ_1 and θ_2.

Moving on now to the data in Table 5.2, the model considered by Chatterjee and Chatterjee (1982) was the following:

$$t_j = \theta_1 + \theta_2 \exp(-j\theta_3) \qquad \theta_2 > 0,\ \theta_3 > 0 \tag{5.16}$$

This model arose from noting that: (1) the times (t_j) are decreasing over the years, (2) the rate of improvement is decreasing, and (3) from physiological considerations it is clear that there must be a lower limit for the times of these events. (In (5.16), θ_1 represents the best attainable time for an event, that is he value corresponding to t_∞.)

able 5.4 Parameter estimates obtained by fitting the model in (5.16) to the data in able 5.2

	100 m	200 m	400 m	800 m
1	0.00 (9.0)	0.00 (16.0)	0.00 (36.0)	87.72 (100.0)
2	10.98 (1.0)	22.45 (4.0)	50.05 (9.0)	30.57 (4.5)
3	0.004 (0.1)	0.005 (0.1)	0.006 (0.1)	−0.030 (0.1)
erations	750	360	410	37

/alues in brackets are the initial parameter values)

Applying first the iterative scheme described by equation (5.14), gives the esults shown in Table 5.4. Clearly there is something odd about the results or the 100, 200, and 400 metre data, and we are forced to the conclusion that or these data the model proposed by Chatterjee and Chatterjee is inappropriate. For the years considered the data give no evidence of a levelling off of e rate of improvement and, consequently no evidence for a lower limit for e times of these events (however reasonable the assumption appears on hysiological grounds). Wootton and Royston (1983) could find no support or any model more complicated than a straight line for these data. The arameter estimates found by fitting the Chatterjee and Chatterjee model to e 800 metre data seem more reasonable although they do not agree with the lues given in the original paper. (They do however agree with the values und by Wootton and Royston.)

Although the parameters in linear and non-linear regression models are ost commonly estimated via least squares using the function in (5.2) as the one to be minimized, there has also been interest in estimation via othe measures of fit, particularly the sum of absolute errors, and the maximun absolute error. These two functions are given by

$$L_1 = \sum_{i=1}^{n} |f_i(y_i : \boldsymbol{\theta})| \qquad \text{(5.17}$$

$$L_\infty = \max\{|f_i(y_i; \boldsymbol{\theta})|\} \qquad \text{(5.18}$$

The interest in these two functions has arisen for a number of reasons, o which the two most important are: (1) many experimental or survey data set do not conform to the Gaussian error assumption required for inference drawn from least squares-fitted regression models, and (2) least square estimates are sensitive to outliers in the data set.

A review of the properties of estimators derived from the minimization o (5.17) or (5.18) is given by Dielman and Pfaffenberger (1982), and variou approaches to deriving such estimators is given in Klingman and Mote (1982).

5.4 LOG-LINEAR AND LINEAR LOGISTIC MODELS

Some of the most useful developments to have taken place in statistics durin the last two decades have concerned the extension of regression models t situations involving categorical variables; examples are log-linear methods logistic regression and Cox's regression model for survival data. Each of thes uses maximum likelihood methods for parameter estimation and in each cas the solution of the maximum likelihood equations involves some type c iterative optimization algorithm. We shall begin by considering the pro cedures used in fitting log-linear models.

5.4.1 Log-linear models

Table 5.5 shows an example of a three-dimensional contingency table. Th analysis of such tables is now routinely undertaken by fitting a series c *log-linear models*, in which the logarithm of the expected frequencies expressed as a linear function of parameters representing different types c association between the variables. The aim is to find the 'simplest' mod which provides an adequate fit to the data. For any such table a 'saturate log-linear model exists which fits the data perfectly since it contains as ma free parameters as cells in the table. For example, the saturated model for three-dimensional table is as follows:

$$\log_e m_{ijk} = u + u_{1(i)} + u_{2(j)} + u_{3(k)} + u_{12(ij)} + u_{13(ik)} + u_{23(jk)} + u_{123(ijk)} \qquad \text{(5.1}$$

where m_{ijk} is the expected frequency in cell ijk of the table and the 'u-term

Table 5.5 Parole violaters

	Number of violaters	Number of nonviolaters
Without a previous criminal record		
Lone offenders	26	27
Group offenders	28	93
With a previous criminal record		
Lone offenders	21	26
Group offenders	36	49

are parameters representing the effects of each variable, each pair of variables, and a three-variable effect. (The parameters are subject to a number of constraints; for details and a full account of log-linear models see Bishop, Fienberg and Holland, 1975.)

By setting some of the parameters in (5.19) to zero, a series of simpler log-linear models may be obtained, each of which may be assessed as a description of the data by estimating the frequencies to be expected under the model and comparing these with the observed frequencies via some goodness-of-fit statistic. The frequencies to be expected under any particular model are estimated by maximum likelihood methods. Birch (1963), shows that the relevant likelihood function is proportional to

$$\sum_{i,j,k} n_{ijk} \log_e m_{ijk} \qquad (5.20)$$

(the exact form of the likelihood function depends on the method of sampling by which the contingency table arose; see Bishop *et al.*, 1975).

By substituting for m_{ijk} from (5.19) or from whatever simpler model is being considered it is easy to find the *sufficient statistics* (see Kendall and Stuart, Volume 2, 1979) for the model. For example, the complete independence model

$$\log m_{ijk} = u + u_{1(i)} + u_{2(j)} + u_{3(k)} \qquad (5.21)$$

has sufficient statistics, $n_{i..}$, $n_{.j.}$ and $n_{..k}$ (for details, again see Bishop *et al.*, 1975). Knowing the sufficient statistics for the model enables maximum likelihood estimates to be found using two results due to Birch, who showed firstly that the sums of estimated frequencies corresponding to the sufficient statistics must equal the sufficient statistics, and secondly that there is a unique set of elementary cell estimates that satisfy both the conditions of the model and these marginal constraints. For example, for the complete independence model there exists a set of estimates $\{\hat{m}_{ijk}\}$ which are such that

$$\hat{m}_{i..} = n_{i..},$$
$$\hat{m}_{.j.} = n_{.j.},$$
$$\hat{m}_{..k} = n_{..k} \tag{5.22}$$

and

$$\hat{m}_{ijk} = \frac{\hat{m}_{i..}\hat{m}_{.j.}\hat{m}_{..k}}{N^2} \tag{5.23}$$

where N is the total number of observations in the table; condition (5.23) is that required for the complete independence of the three variables.

In the case of the three-dimensional table this approach leads to closed-form estimators for all models except that of no three-variable effect. For example, the maximum likelihood estimates of the expected cell frequencies under the complete independence model are

$$\hat{m}_{ijk} = \frac{n_{i..}n_{.j.}n_{..k}}{N^2} \tag{5.24}$$

However, for the log-linear model corresponding to (5.19) with the parameters $u_{123(ijk)}$ set to zero, explicit expressions for the maximum likelihood estimates do not exist and they have to be obtained by some type of iterative scheme. Bartlett (1935) considered a $2 \times 2 \times 2$ table and devised a scheme for obtaining maximum likelihood estimates of all frequencies under the model of no three-variable effect. For such a model the sufficient statistics are $n_{ij.}$, $n_{i.k}$, and $n_{.jk}$ and the condition imposed on the expected frequencies by the model can be shown to be

$$\frac{m_{111}m_{221}}{m_{121}m_{211}} = \frac{m_{112}m_{222}}{m_{122}m_{212}} \tag{5.25}$$

Bartlett's proposal was to set $\hat{m}_{111} = n_{111} + \theta$, so that $\hat{m}_{112} = n_{112} - \theta$ (since $\hat{m}_{11.}$ and $n_{11.}$ have to be equal). The condition (5.25) can then be written in terms of the observed frequencies and the unknown deviation θ as

$$\frac{(n_{111}+\theta)(n_{221}+\theta)}{(n_{121}-\theta)(n_{211}+\theta)}\,\frac{(n_{122}+\theta)(n_{212}+\theta)}{(n_{112}-\theta)(n_{222}-\theta)} = 1 \tag{5.26}$$

This leads to a cubic equation in θ which may be solved in the usual way. Bartlett's approach was later extended to multiple categories, and several alternative procedures were also suggested. For example, Roy and Kastenbaum (1956) used the method of steepest descent supplemented by some numerical graphical procedures. The most usual way of finding maximum likelihood estimates is, however, by using what is known as the *iterative proportional fitting algorithm*, first suggested by Deming and Stephan (1940) in the context of adjusting the observed frequencies in a table to fit a set of marginal totals derived from another source. For a three-dimensional table the method operates by adjusting a set of preliminary estimates to fit successively the appropriate sufficient statistics. For example, for the model of no three-variable effects, the initial estimates are adjusted to fit first, $n_{ij.}$, then $n_{i.j}$, and then $n_{.jk}$ by the steps

$$\hat{m}_{ijk}^{(1)} = \frac{\hat{m}_{ijk}^{(0)} n_{ij.}}{\hat{m}_{ij.}^{(0)}} \tag{5.27}$$

$$\hat{m}_{ijk}^{(2)} = \frac{\hat{m}_{ijk}^{(1)} n_{i.k}}{\hat{m}_{i.k}^{(1)}} \tag{5.28}$$

$$\hat{m}_{ijk}^{(3)} = \frac{\hat{m}_{ijk}^{(2)} n_{.jk}}{\hat{m}_{.jk}^{(2)}} \tag{5.29}$$

This three-step cycle is repeated until convergence to the desired accuracy is attained. Bishop *et al.* (1975) show that this algorithm has the following properties:

1. It always converges to the required set of maximum likelihood estimates.
2. A stopping rule may be used that ensures accuracy to any desired degree in the elementary cell estimates, instead of a rule that only ensures accuracy in one or more summary statistics.
3. The estimates only depend on the sufficient statistics, and so no special provision need be made for sporadic cells with no observations.
4. Any set of starting values may be chosen that conforms to the model being fitted.
5. If direct estimates exist, the procedure yields the exact estimates in one cycle.

To illustrate how the iterative proportional fitting algorithm works in practice we shall use it to find estimated expected values for a no three-variable effect model fitted to the data in Table 5.5. As initial estimates we shall take $m_{ijk}^{(0)} = 1$ for every cell. (These initial estimates clearly fit the no three-variable effect model and so satisfy condition 4 above.) The results from the iterative proportional fitting algorithm are shown in Table 5.6.

Haberman (1974) considers a number of other approaches to the numerical evaluation of maximum likelihood estimates in the context of log-linear models, particularly a modified Newton–Raphson algorithm. Haberman shows that this method has, in general, a superior convergence rate to

Table 5.6 First steps of iterative proportional fitting algorithm in fitting no second-order interaction model to the data in Table 5.5

Cycle 1		
	$\hat{m}_{111}^{(1)} = 26.5$,	$\hat{m}_{112}^{(1)} = 26.5$
	$\hat{m}_{121}^{(1)} = 60.5$,	$\hat{m}_{122}^{(1)} = 60.5$
	$\hat{m}_{211}^{(1)} = 23.5$,	$\hat{m}_{212}^{(1)} = 23.5$
	$\hat{m}_{221}^{(1)} = 42.5$,	$\hat{m}_{222}^{(1)} = 42.5$
Cycle 2		
	$\hat{m}_{111}^{(2)} = 16.45$,	$\hat{m}_{112}^{(2)} = 36.55$
	$\hat{m}_{121}^{(2)} = 37.55$,	$\hat{m}_{122}^{(2)} = 83.45$
	$\hat{m}_{211}^{(2)} = 20.29$,	$\hat{m}_{212}^{(2)} = 26.70$
	$\hat{m}_{221}^{(2)} = 36.70$,	$\hat{m}_{222}^{(2)} = 48.29$

Table 5.7 Estimated expected values found from a Newton–Raphson algorithm for the model of no three-factor effect fitted to the data in Table 5.5

	Number of violaters	Number of nonviolaters
Without a previous criminal record		
Lone offenders	21.81	31.19
Group offenders	32.19	88.81
With a previous criminal record		
Lone offenders	25.19	21.81
Group offenders	31.81	53.19

iterative proportional fitting although the latter is easier to apply to complex models. Haberman applies his Newton–Raphson algorithm to fit a model with no three-factor effect to the data in Table 5.5 and finds the estimated frequencies shown in Table 5.7 after four iterations.

5.4.2 Logistic regression

Many investigations, particularly in medicine, involve a response variable which is binary and we often wish to model the dependence of this variable on a number of explanatory variables. For example, Table 5.8 shows the results of a clinical trial of a particular drug in which we would like to assess the dependence of improvement on the age of a patient and on their weight. One way to model the dependence would be to assume that the probability of the

Table 5.8 Results from a clinical trial

	y_i	Age (years)	Weight (pounds)
1	1	27	140
2	1	35	145
3	1	33	151
4	1	38	138
5	1	23	160
6	1	41	165
7	1	36	135
8	0	34	142
9	0	38	143
10	0	39	170
11	0	37	165
12	0	28	164
13	0	42	169
14	0	44	151
15	0	48	180
16	0	29	167
17	0	39	133
18	0	39	171
19	0	44	171
20	0	47	163

y_i = 1 if patient i improved
= 0 if no improvement or deterioration

dependent variable, y, taking the value 1 say, is a linear function of the explanatory variables

$$P_i = P(y_i = 1) = \sum_{j=1}^{p} \beta_j x_{ij} \qquad (5.30)$$

where y_i is the value of the response variable for subject i and x_{ij}, $j = 1, \ldots, p$ are the values of the p explanatory variables for this subject.

The model in (5.30) is clearly equivalent to the usual multiple regression model for continuous response variables, and the parameters in the model, $\beta_1, \ldots, \beta_p$, might be estimated by least squares methods. Such an approach would be unsuitable for a variety of reasons; the first is that the y's are clearly not normally distributed and therefore no method of estimation that is linear in the y's will in general be fully efficient. In addition there is the possibility that the least squares estimates of the parameters might lead to a fitted value for the probability that did not satisfy the condition $0 < P_i < 1$. Because of these limitations (5.30) is not really a suitable model in this situation. One which is more appropriate is that suggested by Cox (1970) in which the *logistic transformation* of P_i is expressed as a linear function of the explanatory variables; that is

$$\lambda_i = \log \frac{P_i}{1-P_i} = \sum_{j=1}^{p} \beta_j x_{ij} \qquad (5.31)$$

Corresponding to the values of P_i of 0 and 1, λ_i takes values $-\infty$ and ∞; consequently the problem of fitted values outside their range is now eliminated. In many respects (5.31) is the most useful analogue for dichotomous response variables of the ordinary regression model for normally distributed data.

The fitting of the model is (5.31) to a set of data such as that in Table 5.8 involves the estimation of the p parameters $\beta_1, \ldots, \beta_p$. The most usual method of estimation is maximum likelihood, with the likelihood function taking the form

$$\mathscr{L} = \prod_{i=1}^{n} P_i^{y_i}(1-P_i)^{1-y_i} \qquad (5.32)$$

From (5.31) we have that

$$P_i = \frac{\exp\left(\sum_{j=1}^{p} \beta_j x_{ij}\right)}{1+\exp\left(\sum_{j=1}^{p} \beta_j x_{ij}\right)} \qquad (5.33)$$

so that

$$\mathscr{L} = \frac{\exp\left(\sum_{j=1}^{p} b_j t_j\right)}{\prod_{i=1}^{n}\left[1+\exp\left(\sum_{j=1}^{p} b_j x_{ij}\right)\right]} \qquad (5.34)$$

where

$$t_j = \sum_{i=1}^{n} x_{ij} y_i$$

The log-likelihood function is therefore

$$L = \sum_{j=1}^{p} b_j t_j - \sum_{i=1}^{n} \log_e\left[1+\exp\left(\sum_{j=1}^{p} b_j x_{ij}\right)\right] \qquad (5.35)$$

Maximizing L may be achieved using one of the techniques described in Chapter 3; the most commonly used has been Newton–Raphson for which w need the first and second derivatives of L with respect to the parameters.

$$\frac{\partial L}{\partial \beta_s} = t_s - \sum_{i=1}^{n} \frac{x_{is}\exp\left(\sum_{j=1}^{p} b_j x_{ij}\right)}{1+\exp\left(\sum_{j=1}^{p} b_j x_{ij}\right)} \tag{5.36}$$

$$\frac{\partial^2 L}{\partial \beta_r \partial \beta_s} = -\sum_{i=1}^{n} \frac{x_{ir}x_{is}\exp\left(\sum_{j=1}^{p} b_j x_{ij}\right)}{\left[1+\exp\left(\sum_{j=1}^{p} b_j x_{ij}\right)\right]^2} \tag{5.37}$$

(The computer programs for logistic regression implemented in the packages SAS and BMDP use the Newton–Raphson algorithm for estimation.)

Cox (1970) discusses various ways in which initial estimates of the parameters may be obtained, and also points out that multiple maxima of the likelihood function are unlikely unless there are either very limited data or gross discrepancies with the model. McCullagh and Nelder (1983) suggest that the choice of starting values usually reduces the number of iterations required only very slightly so that the choice of initial estimates is usually not critical.

To illustrate how various numerical optimization methods work for the logistic regression problem we shall use the data shown in Table 5.8. Two optimization algorithms were used, the first a quasi-Newton method in which

Table 5.9 Finding maximum likelihood estimators for the parameters in a regression model fitted to the data in Table 5.8

	Quasi-Newton algorithm	Fletcher–Reeves
1.	Intercept = − 1.032	Intercept = − 1.032
	$\hat{\beta}_1 = -0.158$	$\hat{\beta}_1 = -0.158$
	$\hat{\beta}_2 = -0.075$	$\hat{\beta}_2 = -0.075$
	Log-likelihood = − 9.439	Log-likelihood = − 9.439
	Number of iterations = 73	Number of iterations = 156
2.	Intercept = − 1.032	Intercept = − 1.032
	$\hat{\beta}_1 = -0.158$	$\hat{\beta}_1 = -0.158$
	$\hat{\beta}_2 = -0.075$	$\hat{\beta}_2 = -0.075$
	Log-likelihood = − 9.439	Log-likelihood = − 9.439
	Number of iterations = 123	Number of iterations = 191

1. Starting values: intercept = 1.0, $\beta_1 = 0.1$, $\beta_2 = 0.1$
2. Starting values: intercept = 3.0, $\beta_1 = 1.0$, $\beta_2 = 1.0$

approximations to first and second derivatives were obtained by simply evaluating the log-likelihood function a number of times, and the second the Fletcher–Reeves method described in Chapter 3; for this method the explicit expressions for the first derivatives given in (5.36) were used. The results for a variety of starting values are shown in Table 5.9. (The explanatory variables were measured as deviations from their respective means.)

Each method, for each set of starting values, converges to the same final solution, although convergence is slightly faster for the quasi-Newton method.

An important point to note about the use of quasi-Newton methods in a statistical context is that although they do not require explicit expressions for the second derivatives, they do generally provide a very accurate approximation to the inverse Hessian matrix at the optimum. This serves as a useful approximation to the variance–covariance matrix of the estimates (see the discussion in Bunday and Kiri, 1987).

5.5 THE GENERALIZED LINEAR MODEL

Log-linear models and logistic regression are both examples of techniques which Nelder and Wedderburn (1972) describe as *generalized linear models* Classical linear regression and probit analysis (see Finney, 1972), are also examples of this class of model. The essential components of a generalized linear model are as follows:

(a) a random variable, y_i, with $E(y_i) = \mu_i$; this is the response variable;
(b) a set of explanatory variables $x_{i1}, \ldots, x_{ip}$, and an associated *linear predictor* η_i given by

$$\eta_i = \sum_{j=1}^{p} \beta_j x_{ij}; \tag{5.38}$$

(c) a *link function*, g, relating the response variable to the explanatory variables

$$\eta_i = g(\mu_i) \tag{5.39}$$

In this type of formulation classical linear models are such that the response variable has a Gaussian distribution, and the appropriate link function is simply the identity function so that

$$E(y_i) = \sum_{j=1}^{p} \beta_j x_{ij} \tag{5.40}$$

Generalized linear models allow the random response variable to have distributions other than Gaussian and the link function may become an nonotonic differentiable function. For example, in dealing with counts the .ppropriate distribution is generally Poisson. Such a distribution has the estriction that $\mu > 0$; consequently an identity link function might lead to roblems since η could be negative. Models based on independence of robabilities associated with the different classifications of cross-classified ata lead naturally to considering multiplicative effects, and this may be xpressed by using a log link function, $\eta = \log(\mu)$ with its inverse $m = e^{\eta}$. Now additive effects contributing to η become multiplicative effects contri-uting to μ. (More details of other distributions and link functions can be ound in McCullagh and Nelder, 1983.)

Formulated in this way a general method can be derived for fitting the nodels based on maximum likelihood estimation and Fisher's scoring lgorithm. McCullagh and Nelder (1983) show how for generalized linear nodels, this approach may be formulated in terms of an *iterative weighted east squares* involving a dependent variable z and weights w defined as ollows:

$$z = \eta + (y - \mu)\frac{d\eta}{d\mu}, \tag{5.41}$$

$$w = \frac{1}{\mathrm{var}(y)}\left(\frac{d\lambda}{d\eta}\right)^2. \tag{5.42}$$

A current estimate of the linear predictor η is used to derive a corres-onding fitted value, λ, from the appropriate link function as $\eta = g(\mu)$. Using ese values the current values of the dependent variable z and the current eights are calculated. These z values are regressed on the explanatory riables using weighted regression to give a new set of parameter estimates; om these a new estimate of the linear predictor can be found. The process is peated until some convergence criterion is satisfied.

This very general approach to fitting linear models is the basis of the GLIM ftware (see Baker and Nelder, 1978). By specifying a suitable link function, d a suitable distribution, parameter estimates may be found for log-linear

ble 5.10 The application of GLIM to the data in Tables 5.5 and 5.8

ble 5.5	Poisson error, log link: model of no three-variable effect
	Estimated expected values as in Table 5.7; fitting algorithm took 3 iterations to converge
ble 5.8	Binomial error, logistic link: parameter estimates as in Table 5.9. Fitting algorithm took 4 iterations to converge

and logistic models, and Whitehead (1980) and Aitkin and Clayton (198
show how GLIM may be used to find parameter estimates for Cox's regre
sion model. As an example of the use of GLIM, Table 5.10 shows the resul
of analysing the data sets given in Tables 5.5 and 5.8.

McCullagh and Nelder (1983) and Wedderburn (1974) discuss the co
vergence properties of the GLIM fitting algorithm under a variety
circumstances.

5.6 SUMMARY

Many models of practical interest are non-linear rather than linear in t
parameters. Least squares estimation of the parameters in such models m
be attempted by a variety of optimization techniques, including Newto
Raphson, which in this particular case can be arranged into a rather simpl
form.

In addition the analysis of complex data by log-linear models, logis
regression and Cox's regression has now become routine. Parameter esti
ation in each case is via some form of iterative fitting algorithm. Each of the
regression models is included in the general approach of Nelder a
Wedderburn (1972) which is the basis of the GLIM package for generaliz
linear models.

6

)ptimization in multivariate ınalysis

1 INTRODUCTION

Iultivariate analysis deals with data where there are observations on more ıan one variable for each subject or object under investigation, and where ere is some inherent interdependence between the variables. Many of the chniques of multivariate analysis involve large amounts of arithmetic, often an attempt to maximize a likelihood function or minimize some goodness- -fit measure, and several methods, originally suggested several decades ;o, have only become of practical significance with the advent of computers ıd improved optimization algorithms. An outstanding example is the use of aximum likelihood methods for factor analysis, which will be described in e next section. Other techniques which will be discussed in this chapter clude the estimation of the parameters in finite mixtures of multivariate ›rmal densities, latent class analysis and multidimensional scaling.

2 MAXIMUM LIKELIHOOD FACTOR ANALYSIS

ıe factor analysis model postulates that the set of observed or *manifest* riables, $x_1,\ldots,x_p$, are linear functions of a number of unobservable *latent* riables or factors plus a residual term; in algebraic terms the model may be itten

$$
\begin{aligned}
x_1 &= \lambda_{11}f_1+\lambda_{12}f_2+\ldots+\lambda_{1k}f_k+u_1,\\
&\;\cdot\\
&\;\cdot\\
&\;\cdot\\
x_p &= \lambda_{p1}f_1+\lambda_{p2}f_2+\ldots+\lambda_{pk}f_k+u_p
\end{aligned}
\tag{6.1}
$$

ıere $f_1,\ldots,f_k$ represent the k latent variables or *common factors* and $\ldots,u_p$ the residual terms. Equations (6.1) may be written more concisely

$$\mathbf{x} = \mathbf{\Lambda f}+\mathbf{u} \tag{6.2}$$

where $\mathbf{f}' = [f_1 \ldots f_k]$, $\mathbf{u}' = [u_1 \ldots u_p]$ and $\mathbf{\Lambda} = [\lambda_{ij}]$ is the $(p \times k)$ matrix factor loadings. Assuming that the residual terms are uncorrelated both wi each other and the latent variables and that the latter are in standardized for with zero means and unit variances, then the model in (6.2) implies that t covariance matrix of the manifest variables, $\mathbf{\Sigma}$, is given by

$$\mathbf{\Sigma} = \mathbf{\Lambda\Phi\Lambda} + \mathbf{\Psi} \quad (6.$$

where $\mathbf{\Psi}$ is a diagonal matrix containing the variances of the residual ter and $\mathbf{\Phi}$ is the matrix of correlations between the common factors. (Often it i model with orthogonal factors which is of interest, in which case $\mathbf{\Phi}$ is set to t appropriate identity matrix.)

If we have a sample of p-dimensional observations, $\mathbf{x}_1 \ldots \mathbf{x}_n$, assumed arise from a multivariate normal distribution, then the elements of the samp covariance matrix, $\mathbf{S}$, follow a Wishart distribution with $(n-1)$ degrees freedom (see, for example, Mardia, Kent and Bibby, 1979). Consequent the log-likelihood corresponding to the information provided by $\mathbf{S}$ neglecting a function of the observations, given by

$$L = -\tfrac{1}{2}(n-1)\{\log_e|\mathbf{\Sigma}| + \mathrm{trace}(\mathbf{S\Sigma}^{-1})\} \quad (6$$

Since the factor analysis model implies that $\mathbf{\Sigma}$ has the form given in (6. the log-likelihood is a function of $\mathbf{\Lambda}$, $\mathbf{\Phi}$ and $\mathbf{\Psi}$. If we assume for simplicity t $\mathbf{\Phi} = \mathbf{I}$ then estimates of $\mathbf{\Lambda}$ and $\mathbf{\Psi}$ will be found by maximizing L. In practic is slightly more convenient to estimate the parameters by minimizing function

$$F(\mathbf{S}, \mathbf{\Sigma}(\mathbf{\Lambda}, \mathbf{\Psi})) = \log_e|\mathbf{\Sigma}| + \mathrm{trace}(\mathbf{S\Sigma}^{-1}) - \log_e|\mathbf{S}| - p \quad (6$$

This is equivalent to maximizing L since L is $-\tfrac{1}{2}(n-1)F$ plus a function of observations only. (A problem in factor analysis is that the model as defi above is not identifiable and conditions have to be imposed on parameter make the estimation procedure possible; most common is to require t $\mathbf{\Lambda}'\mathbf{\Psi}^{-1}\mathbf{\Lambda}$ be diagonal – see Lawley and Maxwell, 1971, for details.)

In finding the minimum of F in (6.5) it is convenient to use a two-st procedure. Firstly F in (6.5) is minimized over $\mathbf{\Lambda}$ for fixed $\mathbf{\Psi}$ and secondly minimize over $\mathbf{\Psi}$. This approach has the advantage that the first minimizat can be carried out analytically, since it can be shown (see Lawley Maxwell, 1971), that for given $\mathbf{\Psi}$, the matrix of factor loadings which is s that $\partial F/\partial\mathbf{\Lambda} = \mathbf{0}$ and also that $\mathbf{\Lambda}'\mathbf{\Psi}^{-1}\mathbf{\Lambda}$ is diagonal is given by

$$\mathbf{\Lambda}_0 = \mathbf{\Psi}^{-1/2}\mathbf{\Omega}(\mathbf{\Theta} - \mathbf{I}) \quad (6$$

here $\mathbf{\Theta}$ is a diagonal matrix containing the k largest eigenvalues, $\theta_1, \ldots, \theta_k$, the matrix $\mathbf{S}^*$ given by

$$\mathbf{S}^* = \mathbf{\Psi}^{-1/2}\mathbf{S}\mathbf{\Psi}^{1/2} \tag{6.7}$$

ıd $\mathbf{\Omega}$ is a $(p \times k)$ matrix consisting of the corresponding standardized genvectors of $\mathbf{S}^*$. Consequently we now have to minimize the following nction with respect to $\mathbf{\Psi}$:

$$f(\mathbf{\Psi}) = \min_{\mathbf{\Lambda}} F(\mathbf{\Lambda}, \mathbf{\Psi}) = F(\mathbf{\Lambda}_0, \mathbf{\Psi}) \tag{6.8}$$

ɔr this minimization we require the diagonal matrix, $\partial f/\partial \mathbf{\Psi}$, whose diagonal ements are the derivatives $\partial f/\partial \psi_i$ $(i = 1, \ldots, p)$. This matrix is simply $F/\partial \mathbf{\Psi}$ when evaluated at $\mathbf{\Lambda} = \mathbf{\Lambda}_0$. Differentiating F with respect to $\mathbf{\Psi}$ we have ee Lawley and Maxwell, 1971),

$$\frac{\partial F}{\partial \mathbf{\Psi}} = \mathrm{diag}[\mathbf{\Sigma}^{-1}(\mathbf{\Sigma} - \mathbf{S})\mathbf{\Sigma}^{-1}] \tag{6.9}$$

ıere diag ($\mathbf{X}$) represents the diagonal matrix formed from $\mathbf{X}$ by replacing all n-diagonal elements of $\mathbf{X}$ by zeros. So we have that

$$\frac{\partial f}{\partial \mathbf{\Psi}} = \mathrm{diag}[\mathbf{\Sigma}_0^{-1}(\mathbf{\Sigma}_0 - \mathbf{S})\mathbf{\Sigma}_0^{-1}] \tag{6.10}$$

wley and Maxwell (1971) show that this may be written as

$$\frac{\partial f}{\partial \mathbf{\Psi}} = \mathrm{diag}[\mathbf{\Psi}^{-1}(\mathbf{\Lambda}_0\mathbf{\Lambda}_0' + \mathbf{\Psi} - \mathbf{S})\mathbf{\Psi}^{-1}] \tag{6.11}$$

tting (6.11) to the null matrix we have

$$\mathrm{diag}(\mathbf{\Lambda}_0\mathbf{\Lambda}_0' + \mathbf{\Psi} - \mathbf{S}) = \mathbf{0} \tag{6.12}$$

ich leads to

$$\mathbf{\Psi} = \mathrm{diag}(\mathbf{S} - \mathbf{\Lambda}_0\mathbf{\Lambda}_0') \tag{6.13}$$

Lawley and Maxwell remark, 'this equation looks temptingly simple', and uggestive of an iterative scheme of the following form beginning from an tial $\mathbf{\Psi}$ supplied by the investigator:

$$\mathbf{\Psi}^{(i+1)} = \mathrm{diag}(\mathbf{S} - \mathbf{\Lambda}_0^{(i)}\mathbf{\Lambda}_0^{(i)\prime}) \tag{6.14}$$

fortunately there is no guarantee that such a procedure will converge, and xwell (1961) sites an example involving 1100 iterations without achieving vergence!

The first really successful method for the maximum likelihood estimation of the parameters in the factor analysis model was developed during the 1960s b Karl Jöreskog; this is described in detail in Jöreskog (1967). The method use is essentially that of Fletcher and Powell (1963), described in Chapter 3, minimize $f(\boldsymbol{\Psi})$ defined in (6.8). To start the iterative procedure we require a initial $\boldsymbol{\Psi}$ and a symmetric and positive definite matrix, $\mathbf{E}$, which will b updated on each iteration and converges to the inverse of the matrix second-order partial derivatives, $\partial^2 f/\partial\psi_i\partial\psi_j$. As initial estimates of the ψ Jöreskog suggests

$$\psi_i = (1 - \tfrac{1}{2}k/p)(1/s^{ii}) \qquad i = 1, \ldots, p \qquad (6.1$$

where s^{ii} is the ith diagonal element of $\mathbf{S}^{-1}$. This choice appears to wo reasonably well in practice.

An initial estimate of $\mathbf{E}$ might be simply the appropriate identity matr although a more powerful choice would be that suggested by Lawley (196 namely $\mathbf{G}^{-1}$ where the elements of $\mathbf{G}$, g_{ij}, are given by

$$g_{ij} = \phi_{ij}^2 \qquad (6.$$

and the ϕ_{ij} are elements of the matrix, $\boldsymbol{\Phi}$, given by

$$\boldsymbol{\Phi} = \boldsymbol{\Psi}^{-1/2}(\mathbf{I} - \boldsymbol{\Omega\Omega}')\boldsymbol{\Psi}^{-1/2} \qquad (6.$$

Lawley shows that asymptotically

$$\partial^2 f/\partial\psi_i\partial\psi_j = \phi_{ij}^2 \qquad (6.$$

Jöreskog (1967) chose the Fletcher–Powell method for minimizing f in (6 in preference to Newton–Raphson since he did not have analytic expressio for the second-order derivatives and suggested that these would be extrem difficult to obtain. Clarke (1970), however, managed to obtain these seco derivatives and suggested the use of the Newton–Raphson method with modification that on the first, and on any subsequent iteration where matrix of exact second derivatives was not positive definite, Lawle approximation be used. He found empirically that the use of this appro mation gives good reductions in the value of the function in early iterati but is comparatively ineffective near the minimum, whereas near minimum the matrix of exact second derivatives is extremely effective reflection of the general comments made about the Newton–Raphson meth in Chapter 3). Clarke compares his Newton–Raphson algorithms w Jöreskog's method on two data sets and shows that the former does conve more rapidly.

Lee and Jennrich (1979) also compare a number of algorithms for e mation in the factor analysis model using both the method of maxim likelihood and weighted least squares. In the maximum likelihood case t

Table 6.1 Correlations among nine psychological variables

	V1	V2	V3	V4	V5	V6	V7	V8	V9
V1	1.00								
V2	0.52	1.00							
V3	0.39	0.48	1.00						
V4	0.47	0.51	0.35	1.00					
V5	0.35	0.42	0.27	0.69	1.00				
V6	0.43	0.46	0.25	0.79	0.68	1.00			
V7	0.58	0.55	0.45	0.44	0.38	0.37	1.00		
V8	0.43	0.28	0.22	0.28	0.15	0.31	0.38	1.00	
V9	0.64	0.64	0.50	0.50	0.41	0.47	0.68	0.47	1.00

Table 6.2 Three-factor solution found by maximum likelihood factor analysis using Newton–Raphson

Variable	$\hat{\lambda}_1$	$\hat{\lambda}_2$	$\hat{\lambda}_3$	$\hat{\boldsymbol{\Psi}}$
1	0.664	0.321	0.074	0.551
2	0.689	0.247	−0.193	0.428
3	0.493	0.302	−0.222	0.617
4	0.837	−0.292	−0.035	0.213
5	0.705	−0.315	−0.153	0.381
6	0.819	−0.377	0.105	0.177
7	0.661	0.396	−0.078	0.400
8	0.458	0.296	0.491	0.462
9	0.766	0.427	−0.012	0.231

emonstrate that the Fisher scoring algorithm converges far more rapidly an that of Fletcher and Powell.

To illustrate the application of maximum likelihood factor analysis we shall se the correlation matrix shown in Table 6.1, which derives from measuring ine psychological variables on 211 normal children. The estimates of the arameters in a three-factor model fitted to these data using the Newton–aphson algorithm as implemented in the BMD-P package are shown in able 6.2. The history of the iterative procedure is given in Table 6.3a. quation (6.15) was used to provide initial values of the elements of $\boldsymbol{\Psi}$. These alues were 0.424, 0.414, 0.566, 0.250, 0.380, 0.267, 0.394, 0.595, 0.290. The stimates found by the Newton–Raphson procedure are identical with those ported in Lawley and Maxwell (1971), found by using the Fletcher–Powell ethod after two iterations of steepest descent.

For interest the analysis was repeated using other initial values for the ements of $\boldsymbol{\Psi}$. In all cases tried the same final values were obtained although

Table 6.3 Newton–Raphson iterations

(a)

Iteration	Likelihood criterion to be minimized (6.8)
1	0.0874
2	0.0420
3	0.0353
4	0.0351
5	0.0350
6	0.0350

(b)

Iteration	Likelihood function to be minimized (6.8)
1	1.9999
2	1.8811
3	0.8532
4	0.7838
5	0.3401
6	0.1872
7	0.1376
8	0.0387
9	0.0351
10	0.0350
11	0.0350

the number of iterations required altered. For example, starting values of 0.
0.1, 0.9, 0.1, 0.1, 0.1, 0.1, 0.9, 0.1, lead to the results shown in Table 6.3b. W
see that even with these very poor initial values there is only a relatively sma
increase in the number of iterations required to give a solution identical t
that reported above.

6.3 CLUSTER ANALYSIS

Cluster analysis techniques are concerned with discovering whether a set
multivariate data consists of distinct groups or clusters of individuals and,
so, to determine both the number of such groups and their characteristics.
large number of techniques are available, many of which are described
Everitt (1980). From a statistical point of view perhaps the most interesti
methods are those based on a *finite mixture distribution* model (see Ever
and Hand, 1981); here we shall consider two such methods, one suitable f
continuous data and one for data where the variables are dichotomous.

6.3.1 Multivariate normal mixtures

If we assume that the data consist of a number of distinct groups of observations and that within each group these observations have a multivariate normal distribution with a particular mean vector and variance–covariance matrix, then the appropriate distribution for the data as a whole is given by

$$f(\mathbf{x}) = \sum_{i=1}^{g} p_i h(\boldsymbol{\mu}_i, \boldsymbol{\Sigma}_i) \tag{6.19}$$

where g is the number of groups; p_i, $i = 1,\ldots,g$ are the proportion of the observations in each group so that

$$\sum_{i=1}^{g} p_i = 1$$

and $h(\boldsymbol{\mu}_i, \Sigma_i)$ is the multivariate normal density with mean vector $\boldsymbol{\mu}_i$ and covariance matrix Σ_i. Using a set of observations $\mathbf{x}_1,\ldots,\mathbf{x}_n$ assumed to arise from (6.19) the problem is to estimate the $\boldsymbol{\mu}_i$, the Σ_i and the mixing proportions, p_i.

In the univariate case the problem has been considered by many authors since Karl Pearson derived estimates for the five parameters in a mixture of two normals using the method of moments (Pearson, 1894). In the multivariate case estimation has involved primarily maximum likelihood methods, the log-likelihood having the form

$$L = \sum_{i=1}^{n} \log_e \left\{ \sum_{j=1}^{g} p_j h(\boldsymbol{\mu}_j, \Sigma_j) \right\} \tag{6.20}$$

Everitt and Hand (1981) show that the equations resulting from setting the derivatives of L with respect to the parameters equal to zero may be written in the form

$$\hat{p}_k = \frac{1}{n} \sum_{i=1}^{n} \hat{P}(k|\mathbf{x}_i) \tag{6.21}$$

$$\hat{\boldsymbol{\mu}}_k = \frac{1}{n\hat{p}_k} \sum_{i=1}^{n} \mathbf{x}_i \hat{P}(k|\mathbf{x}_i) \tag{6.22}$$

$$\hat{\Sigma}_k = \frac{1}{n\hat{p}_k} \sum_{i=1}^{n} (\mathbf{x}_i - \boldsymbol{\mu}_k)(\mathbf{x}_i - \boldsymbol{\mu}_k)' \hat{P}(k|\mathbf{x}_i) \tag{6.23}$$

where

$$\hat{P}(k|\mathbf{x}_i) = \frac{\hat{p}_k h(\hat{\boldsymbol{\mu}}_k, \hat{\Sigma}_k)}{\sum_{j=1}^{g} \hat{p}_j h(\hat{\boldsymbol{\mu}}_j, \hat{\boldsymbol{\Sigma}}_j)} \tag{6.24}$$

is the estimated posterior probability of observation $\mathbf{x}_i$ arising from component k.

Written in this way we can see that the maximum likelihood estimators for the parameters of a mixture of multivariate normals are analogous to those for estimating the parameters of a single multivariate normal except that each sample point is weighted by the posterior probability (6.24). In the extreme case where $\hat{P}(k|\mathbf{x}_i)$ is unity when $\mathbf{x}_i$ is from component k and zero otherwise, then $\hat{p}_k$ is simply the proportion of observations from component k, $\hat{\boldsymbol{\mu}}_k$ is the mean vector of these observations and $\hat{\Sigma}_k$ their covariance matrix. More generally $\hat{P}(k|\mathbf{x}_i)$ is between zero and one and all the samples play some role in the estimates.

Equations (6.21), (6.22) and (6.23) do not of course give explicit estimators of the parameters since these are involved on both the left- and right-hand sides of the equation through $\hat{P}(k|\mathbf{x}_i)$. Written in this way they do however suggest an obvious iterative scheme for deriving estimates; initial values of the p_k, $\boldsymbol{\mu}_k$ and Σ_k are obtained and these are then used to provide first estimates of the $P(k|\mathbf{x}_i)$, which are then used to provide updated values for all the parameters using equations (6.21) to (6.23). This process is repeated until some convergence criterion is satisfied. Such a procedure is essentially an application of the EM algorithm with the estimation of the posterior probabilities constituting the E step and subsequent insertion of the estimates in the right-hand side of equations (6.21), (6.22) and (6.23) being the M step. This approach to the estimation of the parameters in a mixture of multivariate normal densities was initially suggested by Wolfe (1970).

Other algorithms such as Newton–Raphson might be considered but here there would seem to be major practical problems. For example, with $g = 3$ and $p = 5$ there would be a total of 62 parameters to estimate and so the Newton–Raphson algorithm would involve the repeated inversion of a (62×62) matrix. Of course, if Newton–Raphson converges it does so rapidly since it is of second order, whereas the EM algorithm may be very slow (see later). However, if the separation between components is poor, then the performance of the Newton–Raphson algorithm may still be poor. For example, Dick and Bowden (1973) found that for a mixture of two univariate normals with $p = 0.1$, $\mu_1 = 8.0$, $\sigma_1 = 1.0$, $\mu_2 = 7.0$, $\mu_2 = 1.25$ and $n = 400$ Newton–Raphson failed to converge in about half the samples considered.

A problem which has to be dealt with before we can consider any numerical examples is that of singularities in the likelihood surface. The problem may be illustrated by the following simple example. For a two-component univariate normal mixture the likelihood is a function of the five parameters p, μ_1, μ_2, σ_1 and σ_2, $\mathscr{L}(p, \mu_1, \mu_2, \sigma_1, \sigma_2)$ say. Clearly

$$\mathscr{L}(p, x_i, \mu_2, 0, \sigma_2) = \mathscr{L}(p, \mu_1, x_i, \sigma_1, 0) = \infty \qquad i = 1, \ldots, n \qquad (6.25)$$

and so each sample point generates a singularity in the likelihood function. Similarly any pair of points which are sufficiently close together will generate a local maximum, as will triplets, quadruplets, and so on, which are sufficiently close. Consequently the likelihood surface will be littered with a large number of practically irrelevant global maxima. Day (1969) suggests that because of this problem, attention has to be confined to mixtures in which the covariance matrices are constrained in some way, for example by requiring that $\Sigma_1 = \Sigma_2 = \ldots = \Sigma_g$. Such a restriction alters equation (6.23) to

$$\hat{\Sigma} = \frac{1}{n}\sum_{i=1}^{n} \mathbf{x}_i\mathbf{x}_i' - \sum_{k=1}^{g} \hat{p}_k\hat{\boldsymbol{\mu}}_k\hat{\boldsymbol{\mu}}_k' \tag{6.26}$$

where Σ is the assumed common covariance matrix.

The assumption that all components have the same covariance matrix, although realistic in some situations, is obviously rather restrictive and we should perhaps consider in more detail whether a maximum likelihood approach can only be usefully employed on such a subset of normal mixture distributions. Although Day implies that this *is* the case, other authors indicate the contrary. For example, Duda and Hart (1973) find empirically that meaningful maximum likelihood solutions *can* be obtained in the unequal-variances situation if attention is restricted to the largest of the finite local maxima of the likelihood function. Again simulation work by Hosmer (1973, 1974) shows that for reasonable sample sizes and initial values, the iterative maximum likelihood estimators will not converge to parameter values associated with the singularities, although these may present a problem for maximum likelihood estimation when the sample size is small and the components are not well separated. (For an example of an actual set of univariate data troubled by the singularity problem see Murphy and Bolling, 1967.) Such findings seem to indicate that maximum likelihood will, in many cases, lead to useful estimates of the parameters in a normal mixture even when the assumption of equal variance is not made. (See Cox and Hinkley, 1974, Ch. 9, for further discussion of this problem.)

To illustrate the use of the EM algorithm in the context of multivariate normal mixtures 100 observations were generated from the distribution specified by (6.19), with the following characteristics:

$$p_1 = 0.3, p_2 = 0.3, p_3 = 0.4$$

$$\boldsymbol{\mu}_1' = [0.0, 0.0],$$

$$\boldsymbol{\mu}_2' = [1.0, 1.0],$$

$$\boldsymbol{\mu}_3' = [2.0, 2.0]$$

$$\Sigma_1 = \Sigma_2 = \Sigma_3 = \begin{bmatrix} 1.0 & 0.0 \\ 0.0 & 1.0 \end{bmatrix}$$

A number of different initial estimates were used, these being obtained from the results of applying a particular hierarchical clustering method to the data. The results are detailed in Table 6.4. In each case convergence takes place to a different solution indicating perhaps that the log-likelihood surface contains several local maxima. A disturbing feature is that although the log-likelihood values for each solution are similar, some of the parameter estimates are considerably different.

6.3.2 Latent class analysis

A further finite mixture density is the basis of a method known as *latent class analysis* devised originally by psychologists as an alternative to factor analysis for multivariate binary data. We assume that for each individual we have recorded the presence or absence of p signs or symptoms, and that there are g

Table 6.4 The application of the EM algorithm to estimating parameters in mixtures of multivariate normal densities

1. Starting values from Ward's method

			Iteration history	
	Initial values	Final values	Iteration	Log-likelihood
$\hat{p}_1$	0.47	0.46	1	−145.37
$\hat{p}_2$	0.32	0.37	2	−142.11
			3	−141.24
$\hat{\boldsymbol{\mu}}_1$	1.89	1.83	4	−140.86
	2.05	2.04	5	−140.65
			6	−140.52
$\hat{\boldsymbol{\mu}}_2$	−0.45	0.46	7	−140.43
	0.05	−0.24	8	−140.36
			9	−140.31
			10	−140.23
$\hat{\boldsymbol{\mu}}_3$	1.47	0.22	11	−140.23
	0.05	0.85	12	−140.20
			13	−140.17
			14	−139.91
$\hat{\boldsymbol{\Sigma}}$	$\begin{bmatrix} 1.00 & -0.18 \\ -0.18 & 1.00 \end{bmatrix}$	$\begin{bmatrix} 1.00 & -0.01 \\ -0.01 & 1.00 \end{bmatrix}$	15	−139.63
			16	−139.69
			17	−139.44
			18	−139.39
			19	−139.36
			20	−139.34
			21	−139.33

Table 6.4—*continued*

2. Starting values from furthest neighbour

	Initial values	Final values	Iteration history	
			Iteration	Log-likelihood
$\hat{p}_1$	0.50	0.48	1	−141.10
$\hat{p}_2$	0.45	0.47	2	−139.51
			3	−139.19
$\hat{\boldsymbol{\mu}}_1$	1.94	1.81	4	−139.08
	1.93	1.98	5	−139.04
			6	−139.03
			7	−139.02
$\hat{\boldsymbol{\mu}}_2$	0.32	0.45		
	−0.13	−0.08		
$\hat{\boldsymbol{\mu}}_3$	−1.25	−0.77		
	1.70	1.52		
$\hat{\boldsymbol{\Sigma}}$	$\begin{bmatrix} 1.00 & -0.06 \\ -0.06 & 1.00 \end{bmatrix}$	$\begin{bmatrix} 1.00 & 0.05 \\ 0.05 & 1.00 \end{bmatrix}$		

3. Starting values from group average

	Initial values	Final values	Iteration history	
			Iteration	Log-likelihood
p_1	0.48	0.50	1	−143.98
p_2	0.42	0.43	2	−140.79
			3	−140.53
$\hat{\boldsymbol{\mu}}_1$	1.83	1.77	5	−140.40
	2.06	1.98	6	−140.33
			7	−140.28
			8	−140.25
$\hat{\boldsymbol{\mu}}_2$	−0.06	0.23	9	−140.22
	0.11	0.01	10	−140.20
			11	−140.19
			12	−140.18
$\hat{\boldsymbol{\mu}}_3$	2.00	1.06		
	0.45	0.03		
$\hat{\boldsymbol{\Sigma}}$	$\begin{bmatrix} 1.00 & -0.02 \\ -0.02 & 1.00 \end{bmatrix}$	$\begin{bmatrix} 1.00 & -0.04 \\ -0.04 & 1.00 \end{bmatrix}$		

groups of individuals; within groups we assume that the binary variables are independent of one another – this is the so-called *conditional independence* assumption. The parameters of such a model are the mixing proportions of each group, $p_1, p_2, \ldots, p_g$, and a p-dimensional vector for each group, $\boldsymbol{\theta}_1, \ldots, \boldsymbol{\theta}_g$, the elements of which give the probability of a particular variable being present (recorded as unity, say). The finite mixture density corresponding to this model is as follows:

$$f(\mathbf{x}) = \sum_{i=1}^{g} p_i h(\mathbf{x}; \boldsymbol{\theta}_i) \tag{6.27}$$

where $h(\mathbf{x}; \boldsymbol{\theta}_i)$ is a multivariate Bernouilli density of the form

$$h(\mathbf{x}; \boldsymbol{\theta}_i) = \prod_{j=1}^{p} \theta_{ij}^{x_j}(1 - \theta_{ij})^{1-x_j} \tag{6.28}$$

where $\theta_{i1}, \ldots, \theta_{ip}$ are the elements of $\boldsymbol{\theta}_i$ and $x_1, \ldots, x_p$ are the elements of the random vector $\mathbf{x}$. Given a sample of vectors $\mathbf{x}_1, \ldots, \mathbf{x}_n$ assumed to arise from (6.27) the log-likelihood is given by

$$L = \sum_{i=1}^{n} \log_e \left\{ \sum_{j=1}^{g} p_j h(\mathbf{x}_i; \boldsymbol{\theta}_j) \right\} \tag{6.29}$$

Again Everitt and Hand (1981) show that by differentiating (6.27) with respect to the p_i and the elements of $\boldsymbol{\theta}_i$ (remembering that $\Sigma p_i = 1$), and setting the resulting expressions to zero gives the following equations

$$\hat{p}_k = \frac{1}{n} \sum_{i=1}^{n} \hat{P}(k|\mathbf{x}_i) \tag{6.30}$$

$$\hat{\boldsymbol{\theta}}_k = \frac{1}{np_k} \sum_{i=1}^{n} \hat{\mathbf{x}}_i \hat{P}(k|\mathbf{x}_i) \tag{6.31}$$

where $\hat{P}(k|\mathbf{x}_i)$ is again an estimated posterior probability given by

$$\hat{P}(k|\mathbf{x}_i) = \frac{p_k h(\mathbf{x}_i; \hat{\boldsymbol{\theta}}_k)}{\sum_{j=1}^{g} \hat{p}_k h(\mathbf{x}_i; \hat{\boldsymbol{\theta}}_k)} \tag{6.32}$$

As with the case of finite mixtures of multivariate normal densities (6.30) and (6.31) may be used as the basis of an EM algorithm for finding estimates of the p_k and $\boldsymbol{\theta}_k$. Although other algorithms might be considered, the EM algorithm is most widely used in practice, although its convergence rate may be extremely slow (see the examples below). In its favour, however, it has the merits of simplicity and given appropriate initial values will not converge to parameter values outside their range of (0, 1).

Table 6.5 Artificially generated latent class data

Pattern				
x_1	x_2	x_3	x_4	Frequency
0	0	0	0	3
1	0	0	0	12
0	1	0	0	9
1	1	0	0	7
0	0	1	0	6
1	0	1	0	49
0	1	1	0	4
1	1	1	0	12
0	0	0	1	5
1	0	0	1	9
0	1	0	1	41
1	1	0	1	9
0	0	1	1	5
1	0	1	1	16
0	1	1	1	9
1	1	1	1	4

To illustrate the use of the EM algorithm in latent class analysis it was applied to the data shown in Table 6.5, generated from the density defined by (6.27) with the following parameter values:

$$p_1 = 0.5, p_2 = 0.5$$

$$\boldsymbol{\theta}_1' = [0.8 \;\; 0.2 \;\; 0.8 \;\; 0.2]$$

$$\boldsymbol{\theta}_2' = [0.2 \;\; 0.8 \;\; 0.2 \;\; 0.8]$$

Various initial estimates were tried and the results are given in Table 6.6. Here each set of initial values leads to the same final solution, although the number of iterations needed to achieve convergence differs in each case. However, even for the final set of initial values the number of iterations to convergence is not great and the EM algorithm appears to be particularly efficient for the latent class model.

6.4 MULTIDIMENSIONAL SCALING

A frequently encountered type of data, particularly in the behavioural sciences, is the proximity matrix, arising either directly from experiments in which subjects are asked to assess the similarity or dissimilarity of two stimuli,

Table 6.6 The EM algorithm and latent class analysis

	Initial values			Final values	Number of iterations
	(a)	(b)	(c)		
p_1	0.5	0.5	0.9	0.40	
$\boldsymbol{\theta}_1$	$\begin{bmatrix}0.8\\0.2\\0.8\\0.2\end{bmatrix}$	$\begin{bmatrix}0.2\\0.9\\0.9\\0.1\end{bmatrix}$	$\begin{bmatrix}0.2\\0.9\\0.9\\0.1\end{bmatrix}$	$\begin{bmatrix}0.83\\0.11\\0.82\\0.18\end{bmatrix}$	(a) 29 (b) 34 (c) 39
$\boldsymbol{\theta}_2$	$\begin{bmatrix}0.2\\0.8\\0.2\\0.8\end{bmatrix}$	$\begin{bmatrix}0.5\\0.5\\0.5\\0.5\end{bmatrix}$	$\begin{bmatrix}0.5\\0.5\\0.5\\0.5\end{bmatrix}$	$\begin{bmatrix}0.13\\0.80\\0.24\\0.73\end{bmatrix}$	

Table 6.7 Dissimilarity ratings of Second World War politicians by two subjects. Subject 1 in lower triangle. Subject 2 in upper triangle

	1	2	3	4	5	6	7	8	9	10	11	12
1 Hitler		2	7	8	5	9	2	6	8	8	8	9
2 Mussolini	3		8	8	8	9	1	7	9	9	9	9
3 Churchill	4	6		3	5	8	7	2	8	3	5	6
4 Eisenhower	7	8	4		8	7	7	3	8	2	3	8
5 Stalin	3	5	6	8		7	7	5	6	7	9	5
6 Attlee	8	9	3	9	8		9	7	7	4	7	5
7 Franco	3	2	5	7	6	7		5	9	8	8	9
8 De Gaulle	4	4	3	5	6	5	4		6	5	6	5
9 Mao Tse Tung	8	9	8	9	6	9	8	7		8	8	6
10 Truman	9	9	5	4	7	8	8	4	4		4	6
11 Chamberlain	4	5	5	4	7	2	2	5	9	5		8
12 Tito	7	8	2	4	7	8	3	2	4	5	7	

or indirectly as a measure of the correlation or covariance of the two stimuli derived from a set of variable values common to both. Table 6.7 shows an example of two such matrices. The investigator who collects such data is interested primarily in uncovering whatever structure or pattern among the stimuli is implied by the proximity values. One way to achieve this is to represent the structure by a simple geometrical model or picture in which distances (usually Euclidean) match the observed proximities in some way.

Such a representation may be achieved by using one or other *multidimensional scaling* technique.

The essence of such techniques is to find a set of points to represent each stimulus such that the distances between points match in some way the observed proximities, in the sense that large distances will correspond to small observed similarities, or to large observed dissimilarities, etc. The distances will be a function of the coordinate values of each point, and the latter will be determined by minimizing some goodness-of-fit measure between the fitted distances and the observed dissimilarities. For example, we might simply try to find coordinate values so that

$$\sum_{i,j} (d_{ij} - \delta_{ij})^2 \tag{6.33}$$

is minimized, where δ_{ij} is the observed dissimilarity of two stimuli i and j, and d_{ij} is the corresponding Euclidean distance between the points representing these stimuli. This distance is given by

$$d_{ij} = \left[\sum_{k=1}^{d} (x_{ik} - x_{jk})^2 \right]^{1/2} \tag{6.34}$$

where $x_{i1}, \ldots, x_{is}$ give the d coordinates of the point representing stimuli i. (d will have to be determined although in most cases a fit in 2 or 3 dimensions will be of most practical use so that we can 'look' at the derived configuration.)

In general (6.33) is not used directly; instead a further stage is introduced which involves performing some form of regression analysis between the d_{ij} and δ_{ij} and then minimizing a function similar to (6.33) but with δ_{ij} replaced by the fitted value obtained from the regression. (Details of the procedure are given in Everitt and Dunn, 1983.) Of particular importance in the development of multidimensional scaling was the introduction by Kruskal (1964a) of *monotonic regression* giving a method which used only the rank order of the observed dissimilarities. To minimize the resulting goodness-of-fit measure, or *stress* as it is usually known, Kruskal (1964b) used a steepest-descent algorithm, with a step size computed from a formula arrived at by a considerable amount of numerical experimentation. Much of the subsequent literature on multidimensional scaling involved attempts to improve this algorithm; see for example Guttman (1968), and Lingoes and Roskam (1973).

A further major advance in the development of multidimensional scaling techniques took place in the 1970s with the introduction by Ramsay of a number of formal statistical models combined with parameter estimation by maximum likelihood methods. The details are given in a series of papers Ramsay (1977, 1982), but the essential features of the approach are as follows. We assume that we have I stimuli and R subjects, and that the observed dissimilarity between stimuli i and j for subject r is δ_{ijr}. We seek a se of d-dimensional coordinates for each stimulus, $\mathbf{x}_1, \mathbf{x}_2, \ldots, \mathbf{x}_l$ such that th distances d_{ijk} defined as

$$d_{ijr} = \sum_{k=1}^{d} w_{rk}(x_{ik} - x_{jk})^2 \quad (6.3$$

are as close as possible in some sense to the observed dissimilarities. In (6.3 x_{ik} and x_{jk}, $k = 1, \ldots, d$ are the elements of $\mathbf{x}_i$ and $\mathbf{x}_j$ and w_{rk}, $k = 1, \ldots,$ are the 'weights' given by subject r to each dimension. (These weigh essentially model the difference between subjects.)

As mentioned above, attempting to fit the d_{ijr} directly to the δ_{ijr} is n usually a particularly useful procedure. Instead the dissimilarities are tran formed in some way prior to finding distances and hence coordinates whi provide the 'best' fit. In the MULTISCAL method of Ramsay a number transformations are allowed as described below. The transformati problem, however, often becomes considerably simpler after taki logarithms of both the dissimilarities and the distances. Since taki logarithms is order preserving, it is equally valid to fit log (distance) to l (dissimilarity) as it is to work with the original scales. In addition the use logarithms also often simplifies other aspects of the fitting proble Consequently the three types of transformation allowed in MULTISCAL w be described in terms of log dissimilarities.

(a) *Scale:*

$$f_r(\log \delta_{ijr}) = \log \delta_{ijr} + V_r \quad (6.3$$

The constant, V_r, estimated separately for each subject, changes t location or origin of the log dissimilarity data.

(b) *Power:*

$$f_r(\log \delta_{ijr}) = p_r \log \delta_{ijr} + V_r \quad (6.$$

Here there are two constants to estimate for each subject, p_r and V_r. T transformation has the effect of changing both the scale and the location of log dissimilarities.

(c) *Spline:*

$$f_r(\log \delta_{ijr}) = s_r(\log \delta_{ijr}) + V_r \quad (6.$$

where $s_r(\cdot)$ is a monotone spline function.

This is a considerably more powerful and sophisticated transformation th the other two, but also requires more computation time. Ramsay rec mends that it should be reserved for the final stages of an analysis when ot asic decisions have been made. Monotone splines are discussed in detail in chumaker (1981).

To illustrate the estimation procedure used by Ramsay we shall consider he power transformation defined by (6.37). This transformation implies the ollowing model:

$$p_r \log_e \delta_{ijr} + V_r = \log_e d_{ijr} + \epsilon_{ijr} \tag{6.39}$$

here the ϵ_{ijr} are residuals measuring the amount by which the transformed og dissimilarities deviate from the fitted log distances. To employ maximum kelihood methods it is necessary to assume some form of distribution for iese residuals, and, as in most areas of statistics, that of normality might be a easonable place to begin. So Ramsay assumes that the residuals are normally istributed with mean zero and variance σ^2_{ijr}. Consequently $\log_e \delta_{ijr}$ is assumed have the following density function:

$$\log_e \delta_{ijr} \sim \frac{p_r}{\sigma_{ijr}\sqrt{2\pi}} \exp -\frac{1}{2}\left(\frac{\epsilon_{ijr}}{\sigma_{ijr}}\right)^2. \tag{6.40}$$

Before moving on to consider maximum likelihood estimation some ought needs to be given to possible influences on the variance, σ^2_{ijr}, of dgements. Clearly some subjects are likely to be more precise in their dgements than others due to motivation, experience with the task, and basic ill in making similarity assessments. There are, however, also likely to be rious effects that relate to stimuli or stimulus pairs. Subjects may be more miliar with some stimuli than with others and this may be reflected in dgemental precision being greater for pairs involving more familiar stimuli. capture such effects Ramsay suggests first a model in which

$$\sigma^2_{ijr} = \sigma^2_r \gamma_{ij} \tag{6.41}$$

ere σ^2_r is a measure of within-subject variation and γ_{ij} is a within-pair nponent of judgemental variance; this may be broken down further as

$$\gamma^2_{ij} = (a^2_i + a^2_j)/2 \tag{6.42}$$

iere the coefficients a^2_i, $i = 1, \ldots, I$ are subject to the constraint $\Sigma a_i = I$ d are indices of the relative precision of judgement for each stimuli.

Returning to the distribution in (6.40), we see that the log-likelihood for the dissimilarities of a single subject may be written as

$$L_r = \text{constant} - \sum_i \sum_j \sum_r \log \sigma_{ijr} - \frac{1}{2} \sum_i \sum_j \sum_r (\epsilon^2_{ijr}/\sigma^2_{ijr}) + M_r \log_e p_r \tag{6.43}$$

where M_r is the total number of judgements made by this subject. (If a subje judges all pairs of stimuli, $M_r = I(I-1)/2$.) If, for the moment, we assun that γ_{ij} in (6.41) is unity for all i and j, then it is clear that the maximu likelihood estimator of σ_r^2 obtained from (6.43) is simply

$$\sigma_r^2 = \frac{1}{M_r} \sum_{i=1}^{I} \sum_{j=1}^{J} \epsilon_{ijr}^2 \qquad (6.4$$

Consequently given values for p_r, v_r and d_{ijr}, σ_r^2 may be found directl Ramsay considers a number of different approaches to finding an efficient a reliable algorithm for finding estimates of these parameters, and sugge that because of the large number of parameters involved and because t log-likelihood function behaves rather differently with respect to differe subsets of the parameters, it is necessary to optimize the log-likelihood phases. Finding estimates of the coordinates and weights in (6.35) is, general, the most troublesome part of the process; given these, howev many of the other parameters can be computed immediately.

Ramsay finds that the problem with estimating the coordinates arises par because there may be a large number and partly because the log-likelihoo rather less quadratic with respect to these parameters than to others in

Table 6.8 MULTISCAL II applied to the dissimilarity matrices in Table 6.7

$\sigma_{ijr}^2 = \sigma_r^2 \gamma_{ij}$

History of iterative procedure

Iteration	Log-likelihood
1	−148.24
2	−64.32
3	−44.04
4	−36.08
5	−29.62
6	−23.99
7	−19.34
8	−17.20
9	−15.77
10	−14.13
15	−0.011
20	2.731
25	3.47
29	3.81

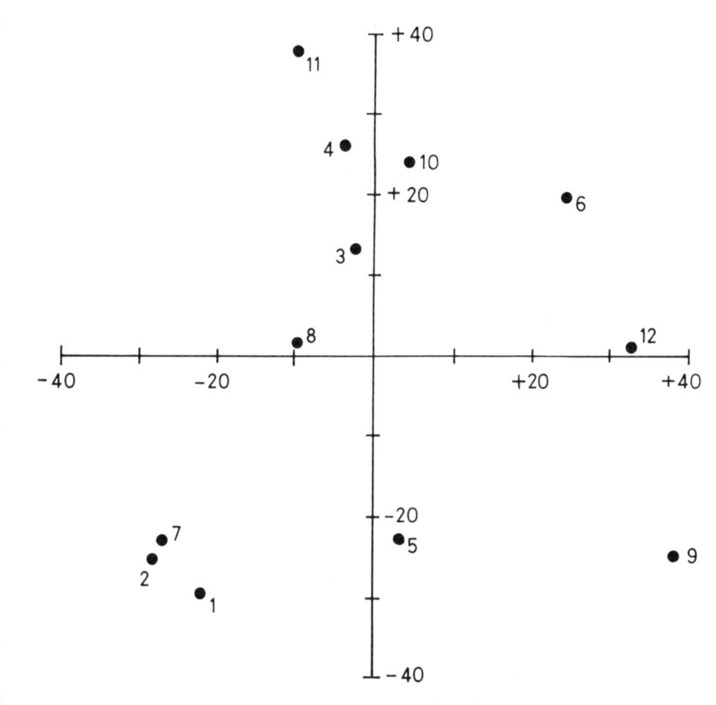

ure 6.1 Two-dimensional configuration obtained from MULTISCAL in the lysis of ratings of dissimilarity of politicians.

del. After trying a number of algorithms Ramsay finally recommends her's scoring procedure (see Chapter 3), and it is this method which is orporated into the MULTISCAL II program. As an example of the eration of this algorithm we shall apply it to the dissimilarity matrices wn in Table 6.7. For this analysis $\sigma_{ijr}^2 = \sigma_r^2 \gamma_{ij}$ was used as a model for the iation in judgements of the politicians. The results of the iterative cedure are outlined in Table 6.8, and the parameter estimates are shown in le 6.9. The two-dimensional configuration obtained is shown in Fig. 6.1.

SUMMARY

advent of computers and the improvement in optimization algorithms has a major influence on the development and the practical application of tivariate methods. Maximum likelihood factor analysis may be carried out tinely, and maximum likelihood methods may also be used in areas such as firmatory factor analysis and structural equation modelling not discussed his chapter (see Everitt, 1984, for details). The EM algorithm may be used erive parameter estimates in finite mixture densities useful as models for cluster analysis. Its convergence can often be rather slow but its simplici makes it more appealing than other possible algorithms. Lastly the method multidimensional scaling has advanced to a stage in which maximum likel hood methods can be used and questions of inference considered. Much of th evolution of this technique has concerned changes in the type of optimizatio algorithm used.

Table 6.9 Parameter estimates obtained from applying MULTISCAL to the dissim larity matrices in Table 6.7

1. Coordinates for two-dimensional solution

	Initial estimates		Final estimates	
	Dimension 1	Dimension 2	Dimension 1	Dimension 2
1	−4.34	−0.29	−22.53	−27.77
2	−5.07	0.37	−28.56	−25.01
3	1.09	1.11	−2.35	13.40
4	2.02	2.74	−3.76	26.16
5	−1.40	−3.33	3.41	−22.72
6	2.69	1.16	24.21	19.87
7	−3.56	0.88	−27.30	−23.03
8	0.12	0.01	−9.73	1.74
9	2.06	−4.87	38.43	−25.34
10	3.93	0.57	4.56	24.02
11	0.35	3.69	−9.16	37.74
12	2.10	−2.05	32.77	0.95

2. Estimates of γ_{ij}

	1	2	3	4	5	6	7	8	9	10	11	12
1												
2	0.01											
3	0.50	0.55										
4	0.01	0.01	0.01									
5	0.58	3.95	0.98	0.32								
6	0.01	0.03	7.18	3.06	0.03							
7	0.01	0.01	0.18	0.35	1.92	0.03						
8	0.50	1.85	6.91	6.20	0.01	0.45	0.36					
9	0.14	0.02	0.01	0.38	0.01	0.10	0.01	2.52				
10	0.05	0.29	0.08	0.97	0.01	0.45	0.01	0.19	0.76			
11	1.15	0.40	0.14	0.01	0.17	2.60	3.27	0.09	1.50	0.01		
12	0.08	0.01	2.17	1.16	1.43	1.49	1.55	5.18	1.51	0.08	0.01	

Table 6.9—*continued*

Estimates of σ_r^2

$\sigma_1^2 = 1.18, \quad \sigma_2^2 = 0.17$

Estimates of weights in (6.35)

Subject	Dimension 1	Dimension 2
	0.96	1.04
	1.04	0.96

Estimates of v_r and p_r in power transformation

	v_r	p_r
Subject 1	0.46	2.37
Subject 2	2.18	1.53

Appendix: exercises

1. Compare a variety of optimization algorithms with a variety of startin values for finding the minimum of each of the following functions.

(a) $f(x_1, x_2) = (16x_1^2 + 16x_2^2 - 8x_1x_2 - 56x_1 - 256x_2 + 991)/15$

$$[\mathbf{x}'_{\min} = (4, 9), f(\mathbf{x}_{\min}) = -18.2]$$

(b) $f(x_1, x_2, x_3) = (x_1 - x_2 + x_3)^2 + (x_1 + x_2 - x_3)^2$

$$[\mathbf{x}'_{\min} = (0, 0, 0), f(\mathbf{x}_{\min}) = 0]$$

(c) $f(x_1, x_2, x_3, x_4) = 100(x_2 - x_1^2)^2 + (1 - x_3)^3 + 10.1(x_2 - 1)^2 + 19.8(x_2 - 1)(x_4 - 1)$

$$[\mathbf{x}'_{\min} = (1, 1, 1, 1), f(\mathbf{x}_{\min}) = 0]$$

2. The probability density function of a mixture distribution composed of tw Poisson components with parameters μ and λ respectively, combined proportions α and $1 - \alpha$ may be written as

$$f(x) = \frac{\alpha e^{-\mu}\mu^x}{x!} + \frac{(1-\alpha)e^{-\lambda}\lambda^x}{x!} \qquad \begin{array}{l} x = 0, 1, \ldots, \infty \\ 0 \le \alpha \le 1. \end{array}$$

Given a sample of observations from this density function find the like hood function, and the likelihood equations for estimating the thr parameters.

Investigate the use of steepest descent, Newton–Raphson and Fletche Reeves in finding the maximum likelihood estimates for the following data.

x	0	1	2	3	4	5	6	7
frequency	830	638	327	137	49	15	3	1

2. A truncated exponential density function has the form

$$f(x) = \lambda e^{-\lambda x}/(1-e^{-\lambda c}) \qquad x \leq c$$

where c is a constant. For a sample of observations $x_1, \ldots, x_n$ from this density show that the likelihood function is given by

$$L = \lambda^n e^{-\lambda} \sum_{i=1}^{n} x_i (1-e^{-\lambda c})^{-n}.$$

Suggest a simple graphical method which could be used to find the maximum likelihood estimator of λ. Compare your method with the results obtained from Newton–Raphson for the following data drawn from such a density with $\lambda = 1.0$ and $c = 1.5$.

0.53495	0.33676	0.59671	1.28840	0.28603
0.18338	0.15082	0.42654	0.81693	0.69833
0.93841	0.06836	0.92658	0.39314	0.08227
0.83003	0.05288	0.60613	0.60410	1.21632
0.62962	0.03493	1.48802	0.51257	1.03245
0.15208	0.12643	0.17467	0.41677	0.66373
1.05877	0.24880	0.20516	0.70610	0.01899
0.45714	0.51985	0.73178	0.09021	0.26240
0.24517	0.06909	0.06497	0.57614	0.52521
0.21695	0.01568	0.76230	0.84435	0.33877

4. Compare the estimates given in Table 5.3 for the data in Table 5.1, with those found by using a least absolute deviation criterion of fit. Investigate the use of several different optimization procedures.

5. A method suggested by Sammon (1969) for obtaining a low-dimensional representation of multivariate data is to seek a set of coordinates $x_1, \ldots, x_{p^*}$ for each individual so as to minimize

$$S = \sum (d_{ij} - d_{ij}^*)^2$$

where

$$d_{ij} = \left[\sum_{k=1}^{p} (y_{ik} - y_{jk})^2 \right]^{1/2}$$

and

$$d_{ij}^* = \left[\sum_{k=1}^{p^*} (x_{ik} - x_{jk})^2 \right]^{1/2}$$

are the Euclidean distances between pairs of points for the original data and for the derived coordinates. Construct a computer program for implementing this procedure and use it to find a two-dimensional representation of the five-dimensional observations given in Table A.1.

Table A.1

	1	2	3	4	5
1	1.79052	2.40779	2.30405	3.33947	4.81159
2	0.41870	2.41622	−1.15054	1.60016	3.77954
3	0.38100	2.17836	−1.51034	1.50106	5.23665
4	1.40827	2.54015	−0.82001	1.21235	2.22920
5	0.98923	2.23567	−1.46415	2.25915	1.72468
6	0.72403	2.12971	−1.32712	2.56672	1.66196
7	1.93870	2.27699	−0.63780	2.62517	2.91837
8	−0.54793	1.61704	−1.34134	1.93518	3.66258
9	0.74240	2.01851	−2.35586	1.28895	3.80879
10	1.76550	2.40141	−0.47575	3.00373	1.96689
11	1.16110	−0.91696	1.13762	−0.65443	0.93723
12	2.91133	−0.79694	2.68309	0.73394	1.44418
13	2.50516	0.46141	0.37595	0.06101	0.79193
14	1.86969	1.05876	1.14048	−0.50032	0.74940
15	1.82928	−0.05657	1.21313	0.12843	0.89604
16	2.26915	0.22230	1.84513	−0.15286	0.62941
17	2.06271	−0.64234	1.59496	0.52112	1.66441
18	1.29071	0.84857	1.37032	0.65502	1.46672
19	1.60536	0.84620	2.92455	0.15850	1.00242
20	1.60489	0.10619	1.78494	−0.41721	0.99682

References

Adby, P. R. and Dempster, M. A. H. (1974) *Introduction to Optimization Methods*, Chapman and Hall, London.

Aitkin, M. and Clayton, D. (1980) The fitting of exponential, Weibull and extreme value distributions to complex censored survival data using GLIM. *Applied Statistics*, **29**, 156–63.

Baker, R. J. and Nelder, J. A. (1978) *The GLIM system*, Release 3, *Generalized Linear Interactive Modelling*, Numerical Algorithms Group, Oxford.

Bartlett, M. S. (1935) Contingency table interactions. *J. Roy. Statist. Soc. Suppl.*, **2** 249–52.

Birch, M. W. (1963) Maximum likelihood in three way contingency tables. *J. Roy. Statist. Soc.*, Ser. B, **25**, 220–33.

Bishop, Y. M. M., Fienberg, S. E. and Holland, P. W. (1975) *Discrete Multivariate Analysis*, The MIT Press, Cambridge, Massachusetts.

Box, M. J., Davies, D. and Swann, W. H. (1969) *Non-linear Optimization Techniques*, Oliver and Boyd, Edinburgh.

Bunday, B. D. (1984) *Basic Optimization Methods*, Edward Arnold, London.

Bunday, B. D. and Kiri, V. A. (1987) Maximum likelihood estimation – practical methods of variable metric optimization methods. To be published in *The Statistician*.

Chatterjee, S. and Chatterjee, S. (1982) New lamps for old: an exploratory analysis of running times in Olympic Games. *Applied Statistics*, **31**, 14–22.

Clarke, M. R. B. (1970) A rapidly convergent method for maximum likelihood factor analysis. *Brit. J. Math. Statist. Psychol.*, **23**, 43–52.

Cox, D. R. (1970) *The Analysis of Binary Data.* Chapman and Hall, London.

Cox, D. R. (1972) Regression models and life tables. *J. Roy. Statist. Soc.*, Ser. B, **34**, 187–220.

Cox, D. R. and Hinkley, D. V. (1974) *Theoretical Statistics*, Chapman and Hall, London.

Cox, D. R. and Oakes, D. (1984) *Analysis of Survival Data*, Chapman and Hall, London.

Cressie, N. A. C. and Keightley, D. D. (1981) Analysing data from hormone-receptor assays. *Biometrics*, **37**, 235–49.

Davidon, W. C. (1959) *Variable metric method for minimization*. AEC (US) Research and Development Report No. ANL 5990.

Day, N. E. (1969) Estimating the components of a mixture of normal distributions. *Biometrika*, **56**, 463–74.

Deming, W. E. and Stephan, F. F. (1940) On a least squares adjustment of a sampled frequency table when the expected marginals are known. *Ann. Math. Statist.*, **11**, 427–44.

Dempster, A. P., Laird, N. M. and Rubin, D. B. (1977) Maximum likelihood from incomplete data via the EM algorithm (with discussion). *J. Roy. Stat. Soc.*, Ser. B, **39**, 1–38.

Dennis, J. E. and Schnabel, R. B. (1983) *Numerical Methods for Unconstrained Optimization and Non-linear Equations*, Prentice-Hall, New York.

Dick, N. P. and Bowden, D. C. (1973) Maximum likelihood estimation for mixtures of two normal distributions. *Biometrics*, **29**, 781–90.

Dielman, T. and Pfaffenberger, R. (1982) LAV (least absolute value) estimation in linear regression: a review. *Studies in Management Science*, **19**, 31–52.

Duda, R. and Hart, P. (1973) *Pattern Classification and Scene Analysis*, John Wiley and Sons, New York.

Dunn, G. (1985) Assays for drug, neurotransmitter and hormone receptors. *The Statistician*, **34**, 357–64.

Everitt, B. S. (1980) *Cluster Analysis*, 2nd edn, Gower, London.

Everitt, B. S. (1984) *An Introduction to Latent Variable Models*, Chapman and Hall, London.

Everitt, B. S. (1984) Maximum likelihood estimation of the parameters in a mixture of two univariate normal distributions; a comparison of different algorithms. *The Statistician*, **33**, 205–16.

Everitt, B. S. and Dunn, D. (1983) *Advanced Methods of Data Exploration and Modelling*, Gower, London.

Everitt, B. S. and Hand, D. J. (1981) *Finite Mixture Distributions*, Chapman and Hall, London.

Finney, D. J. (1972) *Probit Analysis*, 3rd edn, Cambridge University Press.

Fletcher, R. (1971) A general quadratic programming algorithm. *J. Inst. Maths. Applics.*, **7**, 76–91.

Fletcher, R. and Powell, M. J. D. (1963) A rapidly convergent descent method for minimization. *The Computer Journal*, **6**, 163–8.

Forsythe, G. E. and Motzkin, T. S. (1951) Asymptotic properties of the optimum gradient method. *American Mathematical Society Bulletin*, **57**, 183.

Guttman, L. (1968) A general technique for finding the smallest coordinate space for a configuration of points. *Psychometrika*, **24**, 37–49.

Haberman, S. J. (1974) *The Analysis of Frequency Data*, Univ. of Chicago Press, Chicago.

Hosmer, D. W. (1973) On the MLE of the parameters of a mixture of two normal distributions when the sample size is small. *Comm. Statist.*, **1**., 217–27.

Hosmer, D. W. (1974) A companion of iterative maximum likelihood estimates of the parameters of a mixture of two normal distributions under three different types of sample. *Biometrics*, **29**, 761–70.

Jöreskog, K. G. (1967) Some contributions to maximum likelihood factor analysis. *Psychometrika*, **43**, 443–77.

Kalbfleisch, J. D. and Prentice, R. L. (1982) *The Statistical Analysis of Failure Time Data*, Wiley, New York.

Kendall, M. C. and Stuart, A. (1979) *The Advanced Theory of Statistics*, Griffin, London.

Klingman, D. and Mote, J. (1982) Generalized network approaches for solving least absolute and Tchebycheff regression problems. *Studies in Management Science*, **19**, 53–66.

Kruskal, J. B. (1964a) Multidimensional scaling by optimizing goodness-of-fit to non-metric hypotheses. *Psychometrika*, **29**, 1–27.
Kruskal, J. B. (1964b) Non-metric multidimensional scaling; a numerical method. *Psychometrika*, **29**, 115–29.
Lawley, D. N. (1940) The estimation of factor loadings by the method of maximum likelihood. *Proc. of the Roy. Soc. of Edinburgh*, **A40**, 64–82.
Lawley, D. N. (1967) Some new results in maximum likelihood factor analysis. *Proc. of the Roy. Soc. of Edinburgh*, **A67**, 256–64.
Lawley, D. N. and Maxwell, A. E. (1971) *Factor Analysis as a Statistical Method*, Butterworths, London.
Lee, S. Y. and Jenrich, R. I. (1979) A study of algorithms for covariance structure analysis with specific comparisons using factor analysis. *Psychometrika*, **44**, 99–113.
Levenberg, K. (1944) A method for the solution of certain nonlinear problems in least squares. *Quart. Appl. Maths.*, **2**, 164–8.
Lingoes, J. C. and Roskam, E. E. (1973) A mathematical and empirical study of two multidimensional scaling algorithms. *Psychometrika*, **38** (supplement).
McCullagh, P. and Nelder, J. A. (1983) *Generalized Linear Models*, Chapman and Hall, London.
Mardia, K. V., Kent, J. T. and Bibby, J. M. (1979) *Multivariate Analysis*, Academic Press, London.
Marquardt, D. W. (1963) An algorithm for least squares estimation of nonlinear parameters. *J. Soc. Indust. Appl. Maths.*, **11**, 431–41.
Maxwell, A. E. (1961) Recent trends in factor analysis. *J. Roy. Statist. Soc.*, Ser. A, **124**, 49–59.
Murphy, E. A. and Bolling, D. R. (1967) Testing of a single locus hypothesis where there is incomplete separation of the phenotypes. *Amer. J. Hum. Genetics*, **9**, 323–34.
Nelder, J. A. and Mead, R. (1965) A simplex method for function minimization, *The Computer Journal*, **7**, 308–13.
Nelder, J. A. and Wedderburn, R. W. M. (1972) Generalized linear models. *J. Roy. Statist Soc.*, Ser. A, **135**, 370–84.
Pearson, K. (1894) Contribution to the mathematical theory of evolution. *Philosophical Transactions A*, **185**, 71–110.
Ramsay, J. O. (1977) Maximum likelihood estimation in multidimensional scaling. *Psychometrika*, **42**, 241–266.
Ramsay, J. O. (1982) Some statistical approaches to multidimensional scaling (with discussion). *J. Roy. Statist. Soc.*, Ser. A. **145**, 285–312.
Rao, S. S. (1979) *Optimization, theory and applications*, Wiley Eastern, New Delhi.
Rosenbrock, H. H. (1960) An automatic method for finding the greatest or least value of a function. *The Computer Journal*, **3**, 175–84.
Roy, S. N. and Kastenbaum, M. A. (1956) On the hypothesis of no 'interaction' in a multi-way contingency table. *Ann. Math. Statist.*, **27**, 749–57.
Sammon, J. W. (1969) A non-linear mapping for data structure analysis. *IEEE Trans. Computers*, **C18**, 401–9.
Schumaker, L. (1981) *Spline Functions Basic Theory*, Wiley, New York.
Spendley, W., Hext, G. R. and Humsworth, F. R. (1962) Sequential applications of simplex designs in optimization and evolutionary operation. *Technometrics*, **4**, 441–61.
Walsh, G. R. (1975) *Methods of Optimization*, John Wiley, London.
Wedderburn, R. W. M. (1974) Generalised linear models specified in terms of constraints. *J. Roy. Statist. Soc.*, Ser. B, **36**, 449–54.

Whitehead, J. (1980) Fitting Cox's regression model to survival data using GLIM. *Applied Statistics*, **29**, 268–75.
Wolfe, J. H. (1970) Pattern clustering by multivariate mixture analysis. *Multiv. Behav. Res.*, **5**, 329–50.
Wootton, R. and Royston, J. P. (1983) New lamps for old. *Applied Statistics*, **32**, 88–9.
Wu, C. F. J. (1983) On the convergence properties of the E.M. algorithm. *Annals of Statistics*, **11**, 95–103.

Index